Coming to Terms with

Timelessness

Coming to Terms with Timelessness

Daoist Time in Comparative Perspective

edited by

Livia Kohn

Three Pines Press
www.threepinespress.com

9 8 7 6 5 4 3 2 1

Printed in the United States of America
This edition is printed on acid-free paper that meets the American National Standard Institute Z39.48 Standard.
Distributed in the United States by Three Pines Press.

Cover Art: "Measuring Flow: Clock over Yin-Yang and Planets." Design by Brent Cochran.

Front Art: Yin-Yang, Seasons, and Dimensons of Time. Design by Françoise Desagnat.

Library of Congress Cataloging-in-Publication Data

Names: Kohn, Livia, 1956- editor.
Title: Coming to terms with timelessness : Daoist time in comparative perspective / edited Livia Kohn.
Description: First. | St Petersburg : Three Pines Press, 2021. | Includes bibliographical references and index.
Identifiers: LCCN 2021037123 | ISBN 9781931483506
Subjects: LCSH: Time--Religious aspects--Taoism. | Taoism--Relations.
Classification: LCC BL1942.85.T56 C66 2021 | DDC 299.5/142--dc23
LC record available at https://lccn.loc.gov/2021037123

Contents

Acknowledgments

This book is the third edited volume of the project on "Dao and Time." As did its predecessors, *Dao and Time: Classical Philosophy* and *Time in Daoist Practice: Cultivation and Calculation*, it emerged from the 13th International Conference on Daoist Studies, "Dao and Time: Personal Cultivation and Spiritual Transformation." Arranged by Robin Wang, the conference took place in June 2019 at Loyola Marymount University in Los Angeles in close proximity to the 17th Triannial Conference of the International Society for the Study of Time, organized by Paul Harris. I am deeply indebted to them both.

This volume focuses particularly on exploring the many modes in which timelessness plays out in Daoist thought and practice, notably as seen under the light of comparison with, and perspectives in, other philosophies and religions, periods, and cultures. It engages with a variety of topical themes, inspirational thinkers, historical periods, and geographical regions. Doing so, the work adds a yet completely different dimension to the understanding of time in Daoism and opens further venues of exploration, academic discussion, and personal insight.

As editor, I am deeply grateful to everyone involved in the project and particularly the contributors. Not only have they created exceptional work but they have also been wonderfully patient with the editing process. I am excited and pleased to be able to offer their work to the scholarly community in this volume.

—Livia Kohn, July 2021

Introduction

The Nature of Timelessness

LIVIA KOHN

Timelessness marks the primordial state at the beginning of the universe and the ultimate goal of the Daoist endeavor: as such it is the mode of time that most preoccupies both thinkers and practitioners. Called atemporality in J. T. Fraser's system of six major temporalities, it denotes the state of the primeval universe that existed in the "pure becoming of cosmic chaos" (1999, 27), the first fractions of a second right after the big bang about fifteen billion years ago. Consisting of nothing but electromagnetic radiation, it was without time in any form and had "no lawful physical processes" (1982, 50), characterizing a world without causation (2010, 19).

A state of elementary unfolding, timelessness is cosmic chaos, a word that originally means "abyss": the primeval emptiness or dark gorge of the universe (Fraser 1999, 60; 2010, 20). It means eternity as much as infinity, an unlimited (vertical) immediate presence and an endless (horizontal) reaching out. While in some ways dynamic, marked by "the perpetual motion of the photon and the ceaseless vibration of the electron" (1986, 13), it also means the complete absence of choice and freedom, a fundamental darkness and immersion. As J. T. Fraser says, "Only what is temporal is open to change and hence to evolution," can transform into higher and subtler levels of complexity (1986, 13).

This pure cosmic form of timelessness relates to Christian notions of God and eternity (Muller 2016, 18), Plato's ideas and unchanging forms (Davies 1995, 24), as well as the Socratic understanding of the destiny of the soul as having "to climb from the dark, the sensory, and the temporal toward the luminous, the intelligible," that is, the atemporal (Fraser 1986, 13; Lucas 2002, 145). In Christian theology it is sometimes interpreted as endless duration, sometimes as being outside time, but always associated with eternal life and the unchanging validity of religious dogmas. In classical Greek philosophy, it is a key attribute of the Heraclitean logos and the unmoved prime mover, signaling the eternal validity of the rules of numbers (Fraser 2007, 75; Smolin 2013, 9). Many religions, moreover, place timelessness, described by Mircea Eliade as *illud tempus* (1958, ix), on a mythical plane and describe it as a completely "other time beyond our actual time, in which miraculous, primordial, mythological, and symbolic events took place or even still take place" (Franz 1981, 221).

Recognizing timelessness as a precosmic "potentiality of transcending time into blessed eternity" (Fraser 1990, 25), seekers of many religions also see it as the ultimate state of existence and envision it in different forms of paradise and otherworldly realms, places and times that go far beyond the ordinary, reaching into transcendence, immortality, and ultimate perfection (Whitrow 1981, 564; Turner 2010, 336). They speak of attaining complete oneness with the divine, classically realized in mystical union, characterized by ineffability, true knowledge, and passivity as well as a sense of immersion and self-transcendence (James 1936, 370-71).

In this idealized experiential dimension, timelessness signals a realm of complete fulfilledness, a world where one always wanted to be, perfect peace or glorious battle. There is no decay or illness, no suffering, no bodily need—only indestructible, healthy, vibrant life. Here humans can find final and true justice, the ultimate immersion in love and realization of perfection. In other words, timelessness in this sense deeply connects to the human need for refuge from the threat of the passage of time and the knowledge of the end of self (Fraser 2007, 75).

Beyond the far beginning of the universe and ultimate goal of existence, timelessness yet also appears in ordinary life. A vividly present moment without progress in immediate connection to the timespace underlying reality and reaching into the depths of the collective unconscious (Franz 1981, 221), it is to time what emptiness is to space: an openness, a gap, a break, a way of nonbeing. As absolute rest or motionless existence, it offers a pause to the soul from its constant work of self-definition and strife for greater perfection (Fraser 1990, 115).

Classically this is experienced in "a sudden, seemingly spontaneous flash of absolute power or ecstasy" (Ellwood 1980, 69), often called a mystical experience. It is overwhelming, ineffable, and transtemporal, yet full of knowing certainty. Ineffable, it cannot be described in human terms, but descriptions commonly include notions of timelessness and the sense of being grasped by something higher and greater, merged with a state both at the deep root of all existence and its ultimate goal (Kohn 2021, 254).

In other words, people in the grip of a mystical experience have a sense of being outside time and beyond the self, of being grasped by something greater and vaster, and of merging into an immeasurably larger power. They feel as if going beyond all known reality of time and space, as if their ordinary senses have stopped functioning, while yet being in touch with the innermost secrets of the universe (Happold 1970, 45).

The same altered state of consciousness, the overwhelming sense of being fully present in a powerful and empowering way appears also in enlightenment moments in Zen (Suzuki 1956), in the peak experiences explored by Abraham Maslow (1964), in flow as defined by Mihalyi Csikszentmihalyi (1992), and more. While these states can be prepared but not planned, there are certain ways that allow the induction of a sense of timelessness. A prominent example is ritual, which provides a time out from personal strife or suffering, from the chaos of history: changeless and timeless, it offers a structured way to step out of ordinary time (Brand 1999, 43).

This also means that timelessness is a function of consciousness and has a place in the brain. While the left hemisphere provides linear awareness, critical analysis, technological progress, chains of causality, and other organizational structures such as timetables, the right hemisphere functions in modes of wholeness, restoration, contemplation, beauty, cyclicality, and overall integration (Sills 2004, 150; Bogen 1969, 3). It works with overarching patterns and grasps structures all at once, establishing an awareness that reaches beyond the rational, the factual, and the temporal into spheres of balance and harmony, oneness and synthesis. By connecting actively with the right side of the brain, seekers find release from linear temporality and gain a complementary view of the world, engaging in fully holistic perception, that is, "the comprehensive, complete perception of events as intertwined entities, each reciprocally influencing each other" (Sills 2004, 151; Ornstein 1980).

The right hemisphere of the brain is the internal function that opens human beings to the cosmic flow in encompassing timelessness. For example, when Harvard University brain anatomist Jill Bolte Taylor suffered a hemorrhage in her left hemisphere in 1996 and lost its function, she found that "all concepts of time and space evaporated, leaving me instead feeling open-ended, enormous, and expansive" (2009, 68). Thinking in images and resting in the eternal now, her entire self-concept shifted toward fluidity and timeless presence. She says,

> My left hemisphere had been trained to perceive myself as a solid, separate from others. Now, released from that restrictive circuitry, my right hemisphere relished in its attachment to the eternal flow. I was no longer isolated and alone. My soul was as big as the universe and frolicked with glee in a boundless sea. (Taylor 2009, 69)

A number of mind-altering techniques including hypnosis and drugs such as mescaline have the same effect, if less radical and more temporary (Noreika et al. 2014, 529), as do ways of intensifying perception through various rhythmic techniques. "Regularly repeated photic or auditory stimuli tuned to endogenous brain rhythms can evoke altered states of consciousness" (Turner 1986, 242-43): the external rhythm alters or amplifies the endogenous brain rhythm that is necessary for the regulation of neural functions and opens alternate modes of consciousness.

Timelessness being cosmic and ultimate as well as experiential and a function of the brain reinforces its centrality and has led various thinkers to come to the conclusion that the entire paradigm of human temporality is rooted in some sort of monstrous illusion: all time is in fact merely an elaborate product of the human mind. Thus, for example, Lucretius (ca. 95-55 BCE) in *De Rerum Natura* says that "time cannot itself exist." Saint Augustine (354-430) notes that time depends entirely on the mind with its power of distending into past and future through memory and anticipation and as an objective phenomenon remains inexplicable. And Angelus Silesius (1624-1677) notes, "Time is of your own making; its clock ticks in your head. The moment you stop thought, time too stops dead" (Davies 1995, 23). While this may appear depressing at first sight, it yet opens the way to complete control and offers the possibility of taking charge of time: it places the self in a position of greatness. Rather than a cause for despair or hopelessness with the thought that timed reality is a pure mental construct and ultimately unreal, it offers hope and encouragement toward attaining greater and subtler dimensions of life.

This very much is the position of Daoism. Already the *Daode jing* recognizes that the emergence of ordinary timed reality is necessary as the universe gives rise to something rather than nothing yet, whatever form this reality may take, it always remains part of *creatio continua*, the eternal going out and coming back of Dao (Girardot 2008, 77). While the ultimate goal of Daoists is to realize the timeless condition of nothingness before creation in its ultimate form of immortality, they yet live in sympathy with Dao as it unfolds on earth and relish the constant cyclical interplay between chaos and the re-emergence of the world. They are well aware of the powers of the mind in creating and manipulating time, deeply versed in manifold theories and interpretations of the different modes of temporality, and engaged in a plethora of methods and techniques that allow the modification and reorganization of time. But in that they are not alone. Rather, their endeavors are echoed and matched by thinkers and practitioners of other religions and schools of thought, active in other parts of the world and different periods of time.

This Book

This book explores just how Daoists come to terms with timelessness in all these different dimensions while also undertaking close comparisons with other thinkers, religions, and cultures. It offers presentations of a more theoretical, speculative nature in alternation with those that focus on concrete life situations, presenting in turn discussions on issues of personal perception, philosophical speculation, visual representation, self-cultivation, and meaning in life.

To begin, Steve Taylor provides a summary of the different factors that influence the human experience of time as outlined by psychologists. He identifies four "laws" of psychological time and explains the psychological factors that lie behind them, working from his research into time expansion experiences (TEEs). Such experiences typically occur when a person's normal sense of time slows down or expands significantly. They are associated mainly with accidents and emergencies, but also with mystical, psychedelic, and "in the zone" experiences during sports.

Typically linked with a sense of calmness and well-being as well as with alertness and the opportunity to take preventative action, TEEs are interpreted variously but, as Taylor concludes, tend to indicate a shift out of a normal state of consciousness into a different timeworld. They suggest that human beings' normal experience of time is neither objective nor absolute. It is a psychological construct, generated by the psychological processes and structures of our normal self-system. Most specifically, it relates to the strong boundary of our self and the sense of duality this creates. In spiritual experiences (and most altered states of consciousness in general), this state of duality is transcended, also resulting in a transcendence of fast-flowing linear time.

Echoing this, Mercedes Valmisa asks, "What is a Situation?" She leads us in reflections regarding the ontological status of a situation inspired by two main sources: the *Zhuangzi*—a multifarious compilation from Warring States China (ca. 4th c. BCE)—and José Ortega y Gasset's (1883-1955) *Unas Lecciones de Metafísica* (Some Lessons in Metaphysics)—the transcripts of a course on metaphysics by a Spanish philosopher of the early 20th century. Much as other ontologically subjective entities and events, situations do not preexist the intentional subject: instead, they are created alongside an act of noticing.

In Classical Chinese, *shi* 勢, commonly rendered "propensity" and the closest the language comes to our concept of "situation," denotes a dynamic process that incorporates the conscious subjective agent as well as other entities and processes as constitutive elements. Here a situation is not reducible to the discrete phenomena and events that we can discern within a given space-time; rather, it necessitates our thinking about it to arise. These ontological reflections are also important for a philosophy of action. They help us notice the role of attention in the creation of situations—as in the creation of worlds—hence the importance of understanding what the agent notices (Ortega's *reparar*) and fails to notice, what we privilege as worthy of our attention and what passes inadvertent among the world's plural affordances.

The *Zhuangzi* explains that the relational affordances that we actualize and reify as constituting a situation depend on what we are socialized and educated to see when looking at the world, thus situations and agents co-construct one another over time. This acknowledgment is crucial to retrain our agency in order to illuminate our own blind spots, overcome our uncritical certainties which generate absolutist tendencies, and move beyond fixed, reduced, and contingent corners from which to interpret the world.

Looking at similar issues from a different perspective, Joseph L. Pratt, in "Time and Space within Daoism's Holistic Worldview," notes that the *Daode jing* offers a holistic explanation of reality, starting from an ultimate reality of Emptiness, called the Dao. The Dao is followed by the One, a totality capable of encompassing all of conventional reality, and then by the Two, the yin-yang dynamic reflecting the basic interplay between the Dao and the One at the next essentially energetic level. From here emerge the Three, allowing for three-dimensional form and a further fundamental yin-yang dynamic between the energetic yin-yang Two and the material Form Three.

This seamless cosmology and metaphysics, ranging from an absolute nothingness to a play of form, explains how consciousness and cognition exist as the higher energetic Two in relation to form as the lower material Three. It also shows how time and space are a function of form, cognition and, most importantly, consciousness. Finally, Daoism explains how time and space have both yin and yang aspects. The yin circular aspect, the here and now, allows for a direct or immediate path to the transcendent One and ultimately to the Dao. The yang linear aspect, expressed in the standard one temporal and three spatial dimensions, is necessary for the yin circular aspect and, ultimately, for the experience of the Dao to be meaningful.

As this holistic explanation demonstrates, people can achieve a harmonious state, in Daoism called "effortless action" (*wuwei*) and in modern psychology described as state of "flow," where form, time, and space are experienced fully. This transcendent return to the Dao figures in all of life's endeavors from walking a dog to running a business, and can be cultivated by mindfulness practices such as meditation, taiji quan, and yoga.

Equally focusing on the *Daode jing*, Andrej Fech, in "Temporality in Laozi and Plotinus," conducts a comparative study of temporal concepts in the writings ascribed to the two eponymous authors. These two bodies of text were created in different intellectual environments and epochs, and yet there are multiple correspondences in their work, including their view of the ineffable One as the origin of the world, their positing of several precosmic stages, as well as their emphasis of the importance of return.

By comparing the temporal implications of these ideas, this study argues that both sources operate with similar sets of temporal metaphors, implying both cyclical and linear motion. They also cohere in viewing time as emerging prior to the completion of physical reality. As for their differences, in Plotinus, eternity refers to the atemporal state of existence that is a hallmark of the intelligible reality prior to the emergence of the hypostasis Soul. It can be realized by a human individual only by means of contemplative return. The noetic grasp of timeless reality has no prolonging effect on one's bodily existence.

Unlike Plotinus' dichotomy of atemporality (eternity) and temporality (time), the relevant discussion in the *Laozi*, which does not posit the existence of an ideal intelligible world, is based on the distinction between long and short temporal intervals. Accordingly, the main principle of the text, the Way, endures due to its ability to perform timely return. Since here the return is not confined to intellectual understanding, but also includes ethical components, any person able to carry it out in their actions can, in principle, achieve longevity.

Remaining in antiquity but looking at more schematic visions of time, Nada M. Sekulic, in "Magic Square and Perfect Sphere: Time in Daoism and Ancient Greece," explores the visual representation of key features of time. Discussing myth, magic, and temporality in ancient cosmological philosophy and religion, she outlines common settings and differences as found in ancient Greece and early Daoism. She focuses in particular on the idea of the perfect sphere as found in Greek cosmology represented by Pythagoras, Parmenides, and Plato, then compares these notions to the explanations ancient Chinese thinkers give of magic squares such as the *Luoshu* (Writ of the Luo [River]) and the cosmology of the *Yijing* (Book of Changes). Common points include similarities to the Pythagorean theorem as well as medical ideas such as humors and phases. The main difference is the overall meaning of the imperfection of cosmos.

Another mode of the visual representation of time is the topic of Wujun Ke's "Daoist Cinematic Temporality and the Taiwanese New

Wave." She shows just how Taiwanese New Wave directors invoke Daoist references in moments of temporal dilation to comment on Taiwanese society during the post-martial law period, a time of rapid neoliberal transformation. Though noted for their use of slowed-down cinematic time, Edward Yang's *Yi Yi* (A One and a Two, 2000) and Tsai Ming-Liang's *Stray Dogs* (*Jiaoyou*, 2013), masterpieces of the Taiwanese New Wave, are not typically discussed in terms of their references to Daoism.

This changes with this essay. It opens a consideration of Daoist cinematic temporality as discourses around durational time are necessary but insufficient to address the spiritual dimensions of time. By resurrecting Daoist thought to intervene in the homogenous, empty time of late capitalist modernity, Tsai and Yang attend to temporality on a cosmological level, emphasizing acceptance of both the mundane repetitions of daily life as well as the inevitability of change and loss. While Edward Yang's *Terrorizers* is perhaps the most well-known representative of the New Wave for its portrayal of urban alienation and postmodernity, Yang's final film *Yi Yi* pays homage to the spiritual dimension of secular life by invoking Zhuangzi's butterfly dream in the context of a grandmother's passing. Similarly, Tsai Ming-Liang is discussed in terms of slow cinema and disposability, yet he directed *Stray Dogs* with a quote by Laozi in mind. Attending to such details allows us to excavate the existence of unruly temporalities and the enduring influence of spiritual thought on secular life.

Remaining in modernity and moving on to connections with both Western and Buddhist thinkers, Patrick Laude, in "Ways of Time: Seeing Dao through Guénon, Teilhard, and Suzuki," points out that Daoism presents a vision of time that is both cyclical and linear. Its philosophies conceive the world as manifesting in cycles of flow and return, whereas its religious visions see it as moving toward a state of harmony called Great Peace.

This raises first the question of the relationship between spiritual transformation and cosmic change, as reflected in the tension between the Daoist sense of degradation resulting from the collapse of spontaneous oneness with Dao and its millenarian tendencies. Hence the second question: Is the religious view of time descending and declining toward destruction, or ascending toward a final apogee, entropic or progressive? Thirdly, we must envisage that spiritual perception and imagination can also be focused on transcending time and reaching eternity in the present, independently from the vicissitudes of history.

The essay explores of some of the metaphysical and spiritual implications of this triadic power of time in light of the works of René Guénon (1886-1951), Pierre Teilhard de Chardin (1881-1955) and D. T. Suzuki

(1870-1966). The relevance of these authors flows not only from the considerable impact their works had on entire generations of intellectuals and spiritual seekers, but also from the way they were able to articulate and reformulate fundamental insights of Christian theology and Asian wisdom traditions in original, provocative, and seminal ways, providing invaluable tools of intellectual and spiritual revitalization.

How this plays out in religious practice is further explored in Qi Song's "Daoist Aspects of Time Perception in Hakuin's Zen Experiences." She begins by outlining the biography of one of the major Zen masters of Tokugawa Japan, Hakuin Eikaku (1686-1769), then shows how he worked with several different modes of time perception in his various Zen experiences. First of all, during the enlightenment experience of *kenshō*, time for him collapsed into a single instant and vanished completely, opening him to the underlying cosmic reality of timelessness. Then again, when suffering from Zen sickness, time in his experience was dilated and extended into long, drawn-out, and painful segments. Those two represent extreme poles that could also be described as heaven and hell, purity and defilement, oneness and division, and so on.

After learning Daoist techniques, however, Hakuin switched to a gentler and more fluid way of practice, leaving these extremes behind and focusing more on the world of stars, nature, and physicality. Working closely with the natural rhythms, hc emphasized their manifestation within the body through breathing and energy guiding, notably in the ocean of *qi* and the lower elixir field. He adapted his body's patterns to natural time structures as manifest both within and without, while opening his mind as pure consciousness to flow through the different parts of the body. In this manner, he established smooth and harmonious movements, finding the true root of enlightenment deep within himself and going beyond time as a historical marker and ancestral agent.

Linking self and culture, mind and body in the context of how time creates meaning in human life is the focus of the contribution by Juan Zhao. "Synchronicity: A Modern Interpretation of Time in the *Yijing*" traces the emergence of Carl Gustav Jung's (1875-1961) concept of "synchronicity," then analyzes its relationship with the *Yijing* [Boo of Changes), pointing out that it provides a good way of interpreting time (*shi*) as found here. Jung's synchronicity echoes the revolution of the understanding of time that formed part of 20th-century Western physics and philosophy. Yet he also had recourse to the ancient Chinese classic.

By working with the notion time as presented in the *Yijing*, Jung could propose his particular take on the new vision of time, leading to the theory of synchronicity. This added a whole new dimension to the prevalent law of causality—a way of looking at things from a more quantum-based perspective, of simultaneity and coincidence. Proposing this, Jung affirmed another dimension of science, integrating and enhancing the modern significance of the *Yijing*. Doing so, he also gave a new life to the ancient concept of time, making it accessible to, and relevant for, the context of modernity. Seeing time from a comparative perspective and in a dialogue between China and the West, Juan Zhao deepens our understanding of time as well as of the overall situation of human life. Flowing smoothly with the river of time, the human mind can become one with the entire world that makes up its environment, thus opening new and varied paths to moral improvement and spiritual cultivation.

Looking at how time plays out in life from a yet different angle, Jeffrey W. Dippmann focuses on "Immortality and Meaning in Life: A Daoist Perspective." He points out that most Western philosophers have focused on the issue of bodily immortality and the power of death that forces us to utilize life to its fullest, knowing that it can end at any moment. An elixir of life is often seen as counterproductive since it would cause life to become a drudgery of relentless indifference and boredom.

In the Daoist tradition, the notion of *xian* involves the attainment of some sort of immortality. *Xian* come in three types: heavenly immortals who have ascended into heaven and occupy a position in the otherworldly bureaucracy, earthly immortals who are ready for ascension but remain on earth, and those who transcend this world through deliverance from the corpse and leave behind a token or substitute. In keeping with the search for meaning in life, the present study primarily focuses on earthly immortals.

By utilizing tales as found in collections of biographies and classical hagiographies, Dippmann shows that there can indeed be meaning in life for those who attain bodily immortality. For example, the *Liexian zhuan* records the tale of Boshi sheng (Master Whitestone) who, even after 2000 years, desired nothing more than a long life on earth. When queried as to why he didn't ascend to heaven, Whitestone replied that he could not imagine enjoying himself as much in heaven as he did on earth. In contrast to the Western concerns over tedium and despair, the Daoist vision celebrates life, and offers the possibility of a continued joyous existence filled with wonder and hope.

Bibliography

Bogen, Joseph E. 1969. "The Other Side of the Brain." *Bulletin of the Los Angeles Neurological Societies* 34.

Brand, Stewart. 1999. *The Clock of the Long Now: Time and Responsibility*. New York: Basic Books.

Csikszentmihalyi, Mihalyi. 1992. *Flow: The Psychology of Happiness*. London: Rider.

Davies, Paul. 1995. *About Time: Einstein's Unfinished Revolution*. New York: Simon & Schuster.

Eliade, Mircea. 1958. *Birth and Rebirth: The Religious Meanings of Initiation in Human Culture*. New York: Harper and Brothers Publishers.

Fraser, J. T. 1982. *The Genesis and Evolution of Time: A Critique of Interpretation in Physics*. Amherst: University of Massachusetts Press.

_____. 1986. "The Problem of Exporting Faust." In *Time, Science and Society in China and the West*, edited by J. T. Fraser, N. Lawrence, and F. C. Haber, 1-20. Amherst: University of Massachusetts Press.

_____. 1987. *Time: The Familiar Stranger*. Amherst: University of Massachusetts Press.

_____. 1999. *Time, Conflict, and Human Values*. Urbana: University of Illinois Press.

_____. 2007. *Time and Time Again: Reports from a Boundary of the Universe*. Leiden: Brill.

_____. 2010. "Founder's Address: Constraining Chaos." In *Time: Limits and Constraints*, edited by Jo Alyson Parker, Paul Harris, and Christian Steineck, 19-36. Leiden: Brill.

Franz, Marie-Louise von. 1981. "Time and Synchronicity in Analytic Psychology." In *The Voices of Time*, edited by J. T. Fraser, 218-32. Amherst: University of Massachusetts Press.

Girardot, Norman. 2008 [1983]. *Myth and Meaning in Early Taoism: The Theme of Chaos (Hundun)*. Dunedin, Fla.: Three Pines Press.

Happold, F. C. 1970. *Mysticism: A Study and an Anthology*. Baltimore: Penguin.

James, William. 1936 [1902]. *The Varieties of Religious Experience*. New York: Modern Library.

Lucas, J. R. 2002. "Time and Religion." In *Time*, edited by Katinka Ridderbos, 143-65. Cambridge: Cambridge University Press.

Maslow, Abraham H. 1964. *Toward a Psychology of Being*. New York: Van Nostrand Reinhold.

Muller, Richard A. 2016. *Now: The Physics of Time*. New York: W. W. Norton.

Noreika, Valdas, Christine M. Falter, and Till M. Wagner. 2014. "Variability of Duration Perception: From Natural and Induced Alterations to Psychiatric Disorders." In Subjective *Time: The Philosophy, Psychology, and Neuroscience of Temporality*, edited by Valtteri Arstila and Dan Lloyd, 529-56. Cambridge, Mass.: MIT Press.

Ornstein, Robert E. 1980. "Two Sides of the Brain." In *Understanding Mysticism*, edited by Richard Woods, 270-85. New York: Image Books.

Sills, Helen. 2004. "The Composer as Prophet in Time and Uncertainty." In *Time and Uncertainty*, edited by Paul Harris and Michael Crawford, 149-59. Leiden: Brill.

Smolin, Lee. 2013. *Time Reborn: From the Crisis in Physics to the Future of the Universe*. New York: Houghton Mifflin Harcourt.

Suzuki, D. T. 1956. *Zen Buddhism: Selected Writings*. New York: Doubleday.

Taylor, Jill Bolte. 2009. *My Stroke of Insight*. New York: Penguin-Plume.

Turner, Frederick. 1986. "Space and Time in Chinese Verse." In *Time, Science and Society in China and the West*, edited by J. T. Fraser, N. Lawrence, and F. C. Haber, 241-51. Amherst: University of Massachusetts Press.

_____. 2010. "How Death Was Invented and What It Is For." In *Time: Limits and Constraints*, edited by Jo Alyson Parker, Paul Harris, and Christian Steineck, 339-38. Leiden: Brill.

Whitrow, G. J. 1981. "Time and the Universe." In *The Voices of Time*, edited by J. T. Fraser, 564-81. Amherst: University of Massachusetts Press.

The Varieties of Temporal Experience

Time Perception in Altered States of Consciousness

STEVE TAYLOR

A few years ago, I had a car crash. I was driving in the middle lane of a motorway, when a truck pulled out from the inside lane and hit the side of our car, spinning us around, and then hitting us again. As soon as the truck hit us, everything seemed to go into slow motion. There was a very long gap between the sound of impact and the beginning of the car's spin. I looked behind and the other cars on the motorway seemed to be moving extremely slowly, almost as if they were stationery.

I felt as though I had a lot of time to observe the whole scene, and to try to regain control of the car. I was surprised by how clear and vivid everything became, and how much detail I was taking in. There was a strange sense of quietness too. We span around for a few seconds, before careering into a crash barrier on the hard shoulder. Then everything seemed to switch back into normal time again. (Luckily, my wife and I were uninjured.)

My altered perception of time during the seconds of the crash is a common experience. Since I published a book about time perception (Taylor 2007), people regularly have sent me accounts of accidents and other moments of sudden shock which bring about an extreme slowing down of time. One woman told me how she rushed to save her children from the dangers of a nearby fire. She reported:

> Time seemed to stop, enabling me to do this. I moved first one child out and handed her over to a girl that came to help, and then I went back and woke up my eldest, scooped up the baby and then my eldest. . . I will never forget the moments of absolute clarity and calmness. It didn't feel like I was even in my own body. Whatever happened, I remain extremely grateful.

Another woman described a horrific experience when two men tried to rape her, telling me: "I was able to defend myself and escape because everything was so slow that I had time to react faster than the men attacking me."

In another example, a person told me how time slowed down when he helped an old lady who lost her balance and was about to fall over:

> Instinctively I reached out and wrapped my left arm around her, but I was off-balance and the weight of her pulled me down too. But then, between that moment and hitting the pavement a split-second later, everything went into slow motion and, in what seemed like a beautifully choreographed series of movements, I twisted myself underneath her so that she fell on top of me and relaxed my body sufficiently to ensure that I didn't break any of my bones upon impact.

I have received similar reports about robberies and assaults, dangerous confrontations with wild animals, plane crashes, and natural disasters. What the examples suggest is that time perception is not a fixed or constant phenomenon. To a large extent, it relates to states of consciousness—or more specifically, to psychological functions and processes that underlie states of consciousness. The main factor in these accident situations may be that they provide a sudden and dramatic shock which "jolts" us out of our normal state of consciousness.

In this article, I would like to summarize the main aspects of human beings' experience of time, including the different time perceptions we have in different situations. After summarizing some of the ways in which psychologists have tried to explain time perception, I will put forward my own theory, with particular reference to altered states of consciousness. It will become clear that what we experience as a "normal" speed of time is simply the result of a "normal" state of consciousness. As I will show later, there are many varieties of altered states of consciousness—e. g., flow, the "zone" experiences of sports people, psychedelic experiences, deep states of meditation—in which our experience of time alters drastically, in a similar way to accidents and emergencies.

Theories of Time Perception

The flexibility of time perception has been a topic of interest for psychologists ever since psychology emerged as a field of study. In his foundational text *The Principles of Psychology* (1950), William James (1842-1910) made a connection between the human experience of time and intensity of perception. James puzzled over the fact that time seems to speed up as we get older, and suggested that this is because, "in youth, we have an absolutely new experience, subjective or objective, every hour of the day. . . but as each passing year converts some of this experience into automatic routine which we hardly note at all, the days smooth themselves out in recollection to contentless units, and the years grow hollow and collapse" (1950, 624). He also suggested that new experiences, such as foreign travel, have a similar time-expanding effect. As he noted, "rapid and interesting travel" results in the same "multitudinous, and long-drawn-out' time perception as childhood" (1950, 624).

Later psychologists have built upon James' ideas and reached similar conclusions in a more rigorous way. A connection between time perception and information processing was established by the American psychologist Robert Ornstein (1969). In a series of experiments, he played tapes to volunteers with various kinds of sound information on them, such as simple clicking sounds and household noises. At the end he asked them to estimate for how long they had listened to the tape and found that when there was more information on the tape (e. g., when there were double the number of clicking noises), the volunteers estimated the time period to be longer.

Ornstein found that this also applied to the complexity of the information. When they were asked to examine different drawings and paintings, the participants with the most complex images estimated the time period to be the longest. Further experiments along similar lines found that time periods are overestimated when they include more variation (Block and Read 1978) and segmentation (Poynter 1983).

There is still a general consensus among psychologists that time perception is strongly linked to information processing. Two contemporary researchers, William Matthews and Warren Meck, have linked this to factors such as perceptual clarity and ease of information-extraction, which lead to "vivid representations and efficient perceptual decision-making" (2016). This further connects to recent research showing that mindfulness meditation can bring about an "overestimation of duration," suggesting an expanded experience of time (Kramer et al. 2013). This applies to short single meditation sessions and is also a long-term effect of regular meditation practice (Berkovich-Ohana et al. 2011; 2012; Droit-Volet et al. 2015). Similarly, David Eagleman and Vani Pariyadath found that novel stimuli are perceived as longer in duration compared with stimuli that are repeated. They argue that this is due to the increased amount of energy needed to process more novel stimuli. "The experience of duration is a signature of the amount of energy expended in representing a stimulus, i. e., the coding efficiency" (2009, 1841).

Other research has linked time perception to such factors as self-reported mood and arousal. Positive affect and high arousal are associated with a swift passage of time, while negative affect and arousal have the reverse effect. Negative states such as boredom, anxiety, and depression seem to slow down time.[1] In one study, researchers found that oncology patients with lower-than-average levels of well-being reported a slower passage of time (Anderson et al. 2007). In duration estimation studies, fear and threat have also been associated with a slowing down of time (Campbell and Bryant 2007).

Conversely, some studies have shown a link between high levels of hedonic well-being and a swift passage of time. As the saying that "time flies when you're having fun" suggests, tasks that are engaging and enjoyable are reported as passing more quickly (Sackett et al. 2010) In a recent study of how people perceived the passage of time during the UK Covid-19 lockdown, Ruth Ogden (2020) found a link between the passage of time and levels of social satisfaction and stress. People with a higher level of social satisfaction, less stress, and a decreased task load reported a swifter passage of time.

As for the actual psychological or neural mechanisms that determine our subjective experience of time, the most widely accepted theory is the "scalar expectancy theory," originally developed by John Gibbon (1977). This suggests that our sense of time passing comes from a kind of "pacemaker" in our brains. This produces regular pulses that are counted up by an "accumulator" and stored in our memory. At the end of a period of time we unconsciously sense how many times the pacemaker has pulsed and so have a rough idea how much time has passed. According to this theory, time perception slows down when more pulses are accumulated within a given period and speeds up when there are fewer pulses. As yet, however, no one has identified the neural substrates of the "pacemaker-accumulator" system, so it remains just a theory.

Dan Zakay (e. g., 2012) has suggested an additional "attentional gate" component to Gibbon's original theory. The attentional gate regulates the number of signals produced by the pacemaker, and becomes wide or narrow in relation to the allocation of attentional resources. According to him, this means that the more intensely we pay attention to time passing, the slower it seems to pass. Thus, time seems to go slowly when we are waiting in queues or traffic jams. On the other hand, when we are relaxing on holiday or sitting in the garden on a Sunday afternoon, time does not seem so important. We do not pay attention to its passing, which means a lower number of signals from the pacemaker.

---

[1] Droit-Volet et al. 2011; Gil and Droit-Volet 2009; Wyrick and Wyrick 1977.

Four Laws of Psychological Time

In my book *Making Time* (Taylor 2007), I put forward four "laws" of psychological time. They are:

- Time seems to speed up as we get older.
- Time seems to go slowly when we are exposed to new environments and experiences; inversely, time goes quickly when we are in familiar environments and have familiar experiences.
- Time seems to speed up in states of absorption; inversely, it seems to slow down in states of non-absorption, e. g., boredom and anxiety.
- Time often passes slowly, or stops altogether, in altered states of consciousness where our normal self-system—including our normal psychological structures and processes—dissolves or when its boundaries become soft.

The first three of these laws can be explained in terms of the same basic factor, that is, the relationship between information processing and time perception. The fourth law stands on its own and relates to different psychological factors.

Why, then, does time seem to speed up as we get older? This phenomenon has been a topic of discussion among philosophers and psychologists for many decades. One popular explanation is the "proportional" theory, based on the fact that, as we get older, each time period constitutes a smaller fraction of our life as a whole. This theory seems to have been first put forward by the French philosopher Paul Janet (1823-1899), who suggested the law that, as William James describes it, "the apparent length of an interval at a given epoch of a man's life is proportional to the total length of the life itself. A child of ten feels a year as one tenth of his whole life—a man of fifty-nine as one fiftieth, the whole life meanwhile apparently preserving a constant length" (1950, 625).

There is some sense to this theory: it offers an explanation for why the speed of time seems to increase so gradually and evenly, with almost mathematical consistency. One problem, however, is that it tries to explain present time purely in terms of past time. The assumption is that we continually experience our lives as a whole and perceive each day, week, month, or year becoming more insignificant in relation to the whole. But we do not live our lives like this. We live in terms of much smaller periods of time, from hour to hour and day to day, dealing with each time period on its own merits, independently of all that has gone before.

In my view, the speeding-up of time we experience is primarily an effect of the relationship between time perception and information processing, as described above. The main reason why time seems to pass so slowly for children is because of the massive amount of perceptual information they take in from their surroundings. Young children appear to live in a completely different world to adults. Their heightened perception means that they are constantly taking in all kinds of details that pass adults by—tiny cracks in windows, little insects crawling across the floor, patterns of sunlight on the carpet. All phenomena seem to be more vivid and fascinating to them, imbued with more presence and is-ness. And all of this information stretches out time for young children.

However, as human beings get older, we lose this intensity of perception, as the world becomes a more familiar place. We "switch off" to the wonder and is-ness of the world around us; gradually we stop paying conscious attention to our surroundings and experiences. There is a process of progressive familiarization that continues throughout our lives, partly due to the decreasing novelty in our lives, and also due to a psychological "desensitizing mechanism" that makes our experience progressively less vivid (Taylor, 2007, 2010). The longer live, the more familiar the world becomes and the less attention we pay to our experiences.

In my view, this is the main reason why time seems to speed up as we get older. The amount of perceptual information we absorb decreases every year, and time seems to pass progressively faster.

The second and third laws of psychological time can also be explained in terms of the relationship between time perception and information processing.

Time seems to slow down when we are exposed to new environments and experiences (the second law) because the unfamiliarity of new experiences allows the processing of much more perceptual information. In unfamiliar surroundings, we switch out of our normal "desensitized" mode of perception. This is one of the reasons why vacations are so enjoyable, and so essential. On vacation, surrounded by unfamiliarity (and in a mode of relaxation), we often experience a state of spontaneous mindfulness and intensely attend to our surroundings. We rekindle the kind of fresh, first-time vision that is common to young children. As the developmental psychologist Alison Gopnik remarks, "As adults when we are faced with the unfamiliar, when we fall in love with someone new, or when we travel to a new place, our consciousness of what is around us and inside us suddenly becomes far more vivid and intense, like children's" (2006, 211). This vivid perception allows us to process more information, which stretches our experience of time.

The main reason why time goes quickly in states of absorption (the third law) is because our attention narrows to one small focus and we block out information from our surroundings. Think of what happens when you are playing a game of chess, dancing or playing music, watching an enthralling film, or reading an engrossing book. Your absorption becomes so intense that you cease to be aware of your surroundings and even of your own self. You enter a state of "flow" (Csikszentmihalyi 1992). You obviously process information from the focus of your absorption (e. g., the film or the book), but this is a very small amount compared to what you would experience in a normal state of diffuse attention.

It is also significant that in these moments, our minds become very quiet. With our attention so intensely focused outside ourselves, the cognitive chatter of the mind quietens, slows down, and fades away. Thus, one of the major sources of information processing—the associational chatter of thoughts—is massively reduced. Significantly, as Mihalyi Csikszentmihalyi has noted, the absence of cognitive chatter, which he calls "psychic entropy," is one of the characteristics of flow and one of the main reasons why the state brings a sense of well-being. In addition, he mentions that a speeded-up sense of time is one of the characteristics of flow. As he puts it, "Often hours seem to pass by in minutes; in general, most people report that time seems to pass much faster" (1992, 56).

On the other hand, time goes slowly in states of boredom or non-absorption because our attention is unoccupied and a large amount of thought-chatter flows through our minds, bringing a massive amount of cognitive information. This applies to other negative states: research has found that time appears to pass slowly, such as anxiety, pain, fear, and threat.[2] In my view, in all of these states, the main factor is the large amount of cognitive information that flows through our consciousness. In states of psychological discomfort, we are unable to focus our attention and therefore cannot enter a state of absorption

Altered States of Consciousness

As previously noted, the fourth law of psychological time is that time often passes slowly, or stops altogether, in unusual states of consciousness where our normal self-system dissolves or its boundaries become soft. This stands apart from the other three laws, since it deals with a much more dramatic and significant type of experience, in which time seems to slow down massively. This returns us to the type of experiences we exam-

---

[2] Droit-Volet et al. 2011; Gil and Droit-Volet 2009; Wyrick and yrick 1977; Anderson et al. 2007; Campbell and Bryant 2007.

ined at the beginning of this chapter, which can be called time expansion experiences (TEEs).

In addition to accidents and emergencies, TEEs occur in a wide range of different situations, such as under the influence of psychedelic substances (Shanon 2001; Bayne and Carter 2018), in states of meditation and mindfulness and other spiritual experiences, as well as in moments of peak performance in sports, sometimes referred to as being "in the zone" (Murphy and White 1995). In such moments, the customary self-system—normal mental structures and processes—no longer functions. There is a shift into a different state, where our experience of reality alters, including our experience of time.

Although there are some altered states that cause time to pass very quickly, such as flow, most altered states seem to feature a dramatic slowing down. Flow is a relatively mild altered state of consciousness. The states we are going to examine now are more dramatic.

I recently conducted a research study on TEEs (Taylor 2020). Following a pilot study of 22 cases specifically linked to accidents, I collected 74 general reports and examined their causes and characteristics. I found that 40 were related to accidents, mostly involving cars, 12 were connected to meditation or spiritual experiences, 7 happened during sports and games, and another 7 related to psychedelic drugs. In addition, there were a few other minor triggers, like traumatic experiences or listening to music.

Most people described their TEEs as very positive. Almost everyone reported a sense of calmness, despite any danger they might have been facing. Most people noted a sense of alertness or heightened awareness. They felt that their slowed-down sense of time gave them the opportunity to take preventative action. They reported rapid and detailed thinking, with plenty of time to make plans and decisions. As one participant who was in a car accident said, "My head was really clear because I seemed to have so much time to think . . . I will always remember how much time I seemed to have to think and work things out" (Taylor 2020, 13). Some people also reported a sense of quietness, as if noise from their surroundings had become muffled.

In many cases, the sense of time expansion was highly dramatic. Seconds seemed to turn into minutes, or time seemed to stop or disappear altogether. One participant had a TEE during a hockey game. He reported, "The play which seemed to last for about ten minutes. . . occurred in the space of about eight seconds" (2020, 12). Another had such an experience when she fell off a horse. "It only lasted a few seconds for me to be thrown from the horse and hit the ground; however, the whole experience seemed to last for minutes" (2020, 12).

Since many participants reported the sense that their TEE enabled them to take preventive action, it is worth considering the possibility that such experiences are a survival mechanism, a kind of adaptive trait our ancestors developed as a way of increasing their chances of survival in dangerous situations. It would certainly have been beneficial for early humans—surrounded by wild animals and dangerous natural phenomena—to develop the ability to slow down their experience of time in emergency situations.

However, one might also argue that this does not explain why TEEs occur in *non*-emergency situations, such as meditation and psychedelics. The idea that TEEs experiences are an adaptive trait incidentally also works against the notion that they are an illusory phenomenon produced by recollection, as has been suggested by Stetson et. al (2007). After all, it is difficult to see any survival advantage in remembering accidents in more detail afterwards.

As for the neuro-physiological processes involved in TEEs, the Finnish philosopher Valtteri Arstila (2012) has suggested that TEES in accidents may be linked to increased levels of norepinephrine in the brain, related to the fight-or-flight response. He has argued that high levels of norepinephrine could account for characteristics of TEEs such as highly focused attention, increased speed and accuracy of responses, and improved clarity of thought.

However, the most common theme of TEEs in accidents and other situations is calmness and a sense of well-being. This does not fit with the fight-or-flight response or higher levels of norepinephrine. The fight-or-flight response involves anxiety and stress, both of which are strikingly absent in most TEEs.

In addition, we have seen that such experiences do not *just* occur in accidents and emergencies but also in non-emergency situations such as sports, psychedelics, meditation, and listening to music. With the exception of sports, none of these are situations where one would expect to find high levels of norepinephrine. In fact, states of moeditation and other relaxed states—for example, listening to classical music—are usually experienced as states of stillness and inner peace, in stark contrast to a fight or flight response. So, all of this argues against a direct causal connection between high norepinephrine levels and TEEs.

Unfortunately, I cannot offer a complete or detailed explanation of TEEs, but it seems clear that they are linked to altered states of consciousness. As noted earlier, our normal time perception is related to our normal state of consciousness. It is produced by the psychological structures and process which operate when we are in a normal state. When we shift into an altered state, our psychological processes change, and our time perception alters. The more dramatic the altered state, the more dramatic the slowing down of time. This is why TEEs are often linked to meditation, spiritual experiences, and psychedelics. Accidents and emergencies clearly have the capacity to bring about a shift into an altered state due to their sheer shock and intensity. They can suddenly jolt us out of normal consciousness.

All of the above leads to the conclusion that our normal experience of time is neither objective nor absolute. It is simply a psychological construct. What we consider a normal state of consciousness is, as James suggests, "but one special type of consciousness, whilst all about it, parted from it by the flimsiest of screens, there lie potential forms of consciousness entirely different" (1985, 388). And this special type of consciousness is associated with a certain special type of time experience, produced by its own psychological structures and processes. As another contemporary researcher of time perception, Marc Wittman (2018) has noted, since our experience of time is closely bound up with our sense of self and our state of consciousness, when we undergo a shift into a different state of consciousness, due to unusual circumstances or triggers, then we shift into a different timeworld.

Time and Spirituality

The relevance of all this to spirituality—including Daoism—is that spiritual practice can be seen as a conscious attempt to cultivate higher states of consciousness, in which time perception alters and often expands. Time expansion may not be the specific aim of cultivation practices, but it is often a consequence.

A dramatically slowed-down experience of time is certainly a feature of many spiritual experiences (Taylor 2010). As noted near the beginning of this essay, studies of the effects of mindfulness meditation have shown that the practice brings about an "overestimation of duration," suggesting an expanded experience of time (Berkovich-Ohana et al. 2011; 2012; Droit-Volet et al. 2015). Beyond time-estimation studies, there are also many subjective reports by mindfulness meditation practitioners that mention a slowing down of time, as if the present moment has expanded (Kabat-Zinn 2005).

In my view, the reason for this is that spiritual practice dismantles the structures of our normal self-system. Since the normal self-system regulates ordinary perception of time, when its structures diminish, any normal perception of time fades away too. In particular, there may be two main aspects of our normal self-system that generate the normal experience of time. The first is a strong sense of duality between the individual and the world, a sense of being enclosed within one's own mental space while the rest of the world is "out there." In mystical experiences—and sometimes in psychedelic and near-death experiences—this duality dissolves, a state associated with the transcendence of linear time. It is therefore reasonable to suggest that a sense of duality is associated with the experience of linear time—that is, a strong sense of time flowing from the past to the present and into the future.

The second aspect of our normal self-system that may produce human beings' normal experience of time is the automatized nature of normal perception, particularly in the midst of familiar experiences and environments. This results in a reduced amount of information processing, leading to a speeding-up our experience of time. As noted earlier, the increasing familiarity that fills our lives as we grow older may be largely responsible for the sense that time speeds up as people age (Taylor 2007).

A more radical interpretation is to suggest that our normal experience of fast-flowing linear time is illusory. If this experience is partly an effect of a sense of duality, it makes sense that, as the sense of duality becomes weaker (that is, as the self-system becomes more permeable and less separate) time should slow down. This is what happens in TEEs, including accidents and emergencies. At a certain point, duality disappears altogether: the self-system becomes so permeable that there is no distinction between self and world. This occurs in mystical, near-death, and some psychedelic experiences. Any sense of linear time dissolves, suggesting that time is specifically a construct of the *boundaries* of the normal self-system, which fades away as duality fades away and disappears as duality disappears.

The transcendence of duality is the main aim of all spiritual practices and paths, including Daoism. And in transcending duality, we also transcend time.

Bibliography

Anderson, Milton, K. Reis-Costa, and J. R. Misanin. 2007. "Effects of September 11th Terrorism Stress on Estimated Duration." *Perceptual and Motor Skills* 104.3: 799-802. https://doi.org/10.2466/pms.104.3.799-802

Arstila, Valtteri. 2012. "Time Slows Down during Accidents." *Frontiers in Psychology* 3. https://www.frontiersin.org/articles/10.3389/fpsyg.2012.00196/full

Bayne, Tim, and O. Carter. 2018. "Dimensions of Consciousness and the Psychedelic state." *Neuroscience of Consciousness* 2018.1:niy008. https://doi.org/10.1093/nc/niy008

Berkovich-Ohana, Aviva., J. Glicksohn, and Amy Goldstein. 2011. "Temporal Cognition Changes Following Mindfulness, but not Transcendental Meditation Practice." In *Fechner Day 2011: Proceedings of the 27th Annual Meeting of the International Society of Psychophysics*, edited by D. Algom, D. Zakay, E. Chajut, E. Shaki, Y. Mama, and V. Shakuf, 27:245-50. https:/proceeding.fechnerday.com/index.php/proceedings/article/view/423

_____. 2012. "Mindfulness-induced Changes in Gamma-band Activity: Implications for the Default Mode Network, Self-reference, and Attention." *Clinical Neurophysiology* 123.4:700-10. https://doi.org/10.1016/j.clinph.2011.07.048

Block, R. A. and M. A. Reed. 1978. "Remembered Duration: Evidence for a Contextual-change Hypothesis." *Journal of Experimental Psychology: Human Learning and Memory* 4.6:656-65.

Campbell, L. A. and R. A. Bryant. 2007. "How Time Flies: A Study of Novice Skydivers." *Behavior Research and Therapy* 45.6:1389-92. https://doi.org/10.1016/j.brat.2006.05.011

Csikszentmihalyi, Mihalyi. 1992. *Flow: The Psychology of Happiness*. London: Rider.

Droit-Volet, Sylvie, M. Fanget, and M. Dambrun. 2015. "Mindfulness Meditation and Relaxation Training Increase Time Sensitivity." *Consciousness and Cognition* 31.1: 86-97. https://doi.org/10.1016/j.concog.2014.10.007

_____, S. L. Fayolle, and S. Gil. 2011. "Emotion and Time Perception: Effects of Film-induced Mood." *Frontiers in Integrative Neuroscience* 5:33. https://doi.org/10.3389/fnint.2011.00033

Eagleman, David M., and Vani Pariyadath. 2009. "Is Subjective Duration a Signature of Coding Efficiency?" *Philosophical Transactions of the Royal Society B: Biological Sciences* 364(1525):1841-51. https://doi.org/10.1098/rstb.2009.0026

Gibbon, John. 1977. "Scalar Expectancy Theory and Weber's Law in Animal Timing." *Psychological Review* 84.3:279-325. doi:10.1037/0033-295X.84.3.279.

Gil, Sandrine, and Sylvie Droit-Volet. 2009. "Time Perception, Depression and Sadness." *Behavioural Processes* 80.2:169-76. https://doi.org/10.1016/j.beproc.2008.11.012

Gopnik, Alison. 2006. *The Philosophical Baby*. London: Rider.

James, William. 1950 [1890]. *The Principles of Psychology*. London: Dover Press.

_____. 1985 [1902]. *The Varieties of Religious Experience*. Hammondsworth: Penguin.

Kabat-Zinn, Jon. 2005. *Coming to Our Senses: Healing Ourselves and the World through Mindfulness*. New York: Hyperion.

Kramer, R., U. Weger, and D. Sharma. 2013. "The Effect of Mindfulness Meditation on Time Perception." *Consciousness and Cognition* 22.3: 846-52. https://doi.org/ 10.1016/j.concog.2013.05.008

Matthews, William J., and Warren. H. Meck. 2016. "Temporal Cognition: Connecting Subjective Time to Perception, Attention, and Memory." *Psychological Bulletin* 142.8: 865-907. http://dx.doi.org/10.1037/bul0000045

Murphy, M., and R. A. White. 1995. *In the Zone: Transcendent Experience in Sports.* Hammondsworth: Penguin.

Ogden, Ruth. 2020. "The Passage of Time during the UK Covid-19 Lockdown." *PLoS One* 15.7. ISSN 1932-6203

Ornstein, Robert. 1969. *On the Experience of Time.* Hammondsworth: Penguin.

Poynter, William D. 1983. "Duration Judgment and the Segmentation of Experience." *Memory and Cognition* 11.1:77-82.

Sackett, Aaron M., Tom Meyvis, Leif. D. Nelson, Benjami A. Converse, and A. L. Sackett. 2010. "You're Having Fun when Time Flies: The Hedonic Consequences of Subjective Time Progression." *Psychological Science* 21.1:111-17. https://doi.org/ 10.1177/0956797609354832 PMID: 20424031

Shanon, Benny. 2001. "Altered Temporality." *Journal of Consciousness Studies* 8.1:35-58.

Stetson, Chess, Matthew P. Fiesta, and David M. Eagleman. 2007. "Does Time Real ly Slow Down During a Frightening Event?" PLoS ONE. www.doi.org/10. 1371/ journal. pone.0001295

Taylor, Steve. 2007. *Making Time: Why Time Seems to Pass at Different Dpeeds and How to Control It.* Cambridge: Icon Books.

_____. 2010. *Waking from Sleep.* London: Hay House.

_____. 2020. "When Seconds Turn into Minutes: Time Expansion Experiences in Altered States of Consciousness." *Journal of Humanistic Psychology* https://doi.org/10.1177/0022167820917484.

Wittman, Marc. 2018. *Altered States of Consciousness: Experiences out of Time and Self.* Cambridge, Mass: MIT Press. https://doi.org/10.7551/mitpress/11468. 001.0001

Wyrick, Robert A., and Lynda C. Wyrick. 1977. "Time Experience during Depression." *Archives of General Psychiatry* 34.12:1441-43. https://doi.org/10.1001/ archpsyc. 977.01770240067005

Zakay, Dan. 2012. "Experiencing Time in Daily Life." *The Psychologist*, August 2012: 578-fd81.

What is a Situation?[1]

MERCEDES VALMISA

Sometimes we "walk into a situation" without even realizing it and do not know how to walk out of it. While we often "encounter a situation" that we must immediately "face," at other times we may be able to sit back and "see how it unfolds" before taking any action. All these idioms, expressive of our conventional understanding, suggest that a situation is something that exists before our encounter with it; something that is out there before we can even notice and acknowledge it; something that, in sum, possesses a separate and independent existence with identity and boundaries of its own.

My analysis of what a situation is leads me away from this essentialist commonsensical understanding toward a view that at first sight might seem counter-intuitive, but which I expect to become self-evident for the reader by the end of this paper. In my analysis, I will be using interpretive keys and concepts from two main sources: the *Zhuangzi* 莊子 and José Ortega y Gasset's (1883-1955) *Unas Lecciones de Metafísica* (Some Lessons in Metaphysics). That is, a multifarious philosophical compilation from the Warring States (ca. 4th c. BCE) and the transcripts of a course on metaphysics by a Spanish philosopher of the early 20th century.[2]

---

[1] I presented an earlier version of this paper at the Scott and Heather Kleiner Colloquium Series in the Philosophy Department at University of Georgia in September 2020. I thank all the participants for the engaged discussion, and the many relevant questions and insights they offered.

[2] The only surviving version of the *Zhuangzi* was edited and annotated by Guo Xiang 郭象 in the 3rd century CE. However, its materials may have originated in the Warring States through the Western Han period (5th c. BCE-1st c. CE). Some scholars point out that the *Zhuangzi* was entirely a product of the court of Liu An 劉安, King of Huainan 淮南 (2nd c. BCE), a hypothesis I find plausible.

Ortega y Gasset taught his course on metaphysics in Madrid in 1932 and 1933 at the tertulia of the *Revista de Occidente*, founded by himself in 1923, which served the purpose of periodical publication and perpetual seminar. His metaphysics lessons were published posthumously in 1966.

Like a Building

Let us start to shape this counter-intuitive view by reflecting on the similarities and differences between a situation and a building: both are entities into which we may walk. If we turn the corner on Springs Avenue toward Bufford Avenue and walk a few meters, we run into the Gettysburg branch of the U.S. Postal Service. According to the substantialist realist account that still guides most of our intuitions about the world—a world composed of individual substances with independent existence, ontologically divided into subjects and objects, knowers and known, minds and matter, etc.—the bricks that make up the space that functionally works as the post office were firmly erected and remain relatively stable at this location regardless of whether anyone visits, sees, or interacts with them.

Is a situation in any respect similar in its spatio-temporal and ontological existence to the post-office building of the substantialist realist account? That is, how does a situation relate to time and space? Is a situation, like a building, something that is happening out there so that I can walk into it? Namely, has a situation an independent reality of its own regardless whether or not someone happens to notice it? Is a situation real in the same way as the post office is real? I will answer this last question both in the positive and the negative—the negative answer being most relevant for this particular discussion.

On the one hand, a situation is like the post-office building insofar both need to be recognized as such—i. e., given meaning—to exist. In John Searle's vocabulary, both the post office and a situation are not "brute facts" but "social facts" (1995, ch. 1). Without people to agree on the identity and function of the post office building, it would just be a collection of wood, cement, metal, clay, and other materials put together in a way that creates empty spaces inside. Likewise, I will argue, a situation is a subject-dependent, ontologically subjective reality, which needs to be acknowledged as such in order to exist.

On the other hand, in the case of the post-office building of the conventional realist account, the concrete bricks and the empty spaces they create in a given location continue to exist no matter whether someone notices them or not. They might not become a post office unless there is social agreement upon it, but the materials themselves necessitate no such recognition in order to exist out there in time and space. In this sense, which is relevant for this discussion, a situation is very different from the post-office building. A situation is not unless someone thinks of it, acknowledges it, interacts with it. Despite the expression, it is not something we can just walk into as if it preexisted our noticing. Rather, a situation is created along with our act of noticing.

Nevertheless, were we to take issue with the conventional account that emerges from a substance metaphysics—one that assumes the existence of individual "things" owning inherent properties and being "out there" in time and space—the post-office building would take on a very different look. If we understand a building through the lenses offered by particular accounts on contemporary quantum physics and the relational ontology systems consistent with them, our answer to the question will differ. For example, in Karen Barad's account (2006), primordially there are not "things" but "phenomena"—neither individual objects nor mental impressions but entangled material agencies, a term that points at the inseparability of the object and the measuring agencies by means of which the object emerges with specific boundaries and properties. Phenomena, as opposed to individual objects, are configurations facilitated by particular practices among different agencies, both human and nonhuman, which both produce and are produced by these practices themselves.[3]

The world is not independent from our experimental exploration of it; nor are we independent from the world. There is no separation between object and subject (building and perceiver of the building) at an ontological level, for both emerge as distinct only as a result of particular practices of observation and measurement.

In this relational non-substantialist account, the post-office building emerges as a result of the specific materializations of which we are part. And yet a building is a real physical entity—though not inherently fixed and delineated in its boundaries and properties until a particular configuration of agencies delimits and determines it. As a phenomenon, a building is both the matter and the actions of measuring, conceptualizing, using, or interpreting (i.e., acting along with) the matter. It is neither inner nor outer but both: a constantly fluid enacting of boundaries.

A situation is much like the building in this relational view, but a crucial difference remains. The post-office building enjoys more ontological stability than a situation; it is objectified and stabilized through obstinate

---

[3] As Barad notes, quantum physics is often romanticized as a less Eurocentric, androcentric, imperializing, etc. theory that saves us from Western essentialism. However, we must acknowledge that only *some* aspects within *specific* accounts of quantum physics may act in this way, helping us challenge the assumed separation between subject and object, human and nonhuman, natural and cultural, freedom and determinism, physics and philosophy, and other binaries that have structured Western thought (2006, 67-68).

and enduring linguistic, sociocultural, and institutional practices in ways that situations rarely do—as for example, through the practices of naming buildings and locating them in maps as points of reference. Situations in contast to buildings are more ontologically fluid and dependent on moments of condensation.

In Classical Chinese, the word that most closely approximates the meaning of situation is *shi* 勢, "propensity" (see Jullien 1995). As Roger Ames notes, *shi* is "an ongoing process that includes agency within it, (and which) means at once 'situation,' 'momentum,' and 'manipulation.' *Shi* includes all of the conditions that collaborate to produce a particular situation, including place, agencies, and actions" (1998, 227). For ancient Chinese philosophers, *shi* was not something external to the agent or an entity the agent might encounter independently. Rather, it was a process that incorporated the conscious subjective agent as well as other entities and processes as constitutive elements. In contrast to its conventional usage in modern languages, a situation as we are going to understand it, is something made out of us as much as we are made out of it.

A situation, then, is not something externally recognizable that we can see from a distance, not in the way that we can see the bricks that constitute a building. Discussing the constitutive power of the gaze, Jean-Luc Marion observes that the look between two people remains invisible, for it does not belong to one pair of eyes nor the other but only exists in the relation between the two. The gaze itself "remains unable to be looked at," and yet it is by means of being given to the other that it appears (2001, 115). In the same way, we can never look at a situation: we can look at agencies, events, phenomena, or processes that are at rest or in movement; we can look at changes that are given to us by their own appearing; we can look at happenings and our own feelings and reactions to them, but the situation itself remains evasive, ungraspable, invisible.

With the above discussion, we are getting at the following key fact: a situation is not identical with the discrete phenomena and events that we can discern in conjunction with the emergence of a given space-time. A situation is composed of certain available spatiotemporal affordances, but it is not reduced to them nor is it the sum of these parts. Those parts are just like the bricks of the post-office building: materials used to flesh out an entity that would not have a determinate meaning, identity, or boundaries without its entanglement with the agent's recognition. A situation belongs with us, the thinking subjective entities who conform and indeed create it by thinking about it.

The Role of Attention

"Seeing" a situation hence is an exercise of introspection where we become aware of our consciousness highlighting particular aspects of the world, providing them with meaning, and reifying them—making them a thing. In this way, seeing a situation is not so much looking outward as it is a form of self-awareness. By looking out and identifying a situation, we are looking in. And simultaneously, by looking in we discover which outward entities have been selected as the focus of our attention (the building's bricks), for awareness and meaning are always of something.

In and out, self and world, agent and situation are interdependent to the extent that no dichotomy can be established between them and no side takes priority over the other despite their obvious asymmetry. After all, it is the subject—a consciousness with intentionality—that creates the situation by endowing it with existence and meaning, by making it a thing over all other possible configurations of relations that lie in the background, available to be picked out by any given consciousness.

A situation only arises in a coexisting mode of being recognized as such by a subjectivity. This action of recognizing is simultaneously an act of creation: it consists of establishing a focus on a fixed set of space-time relations. To further this analysis, we want to avail ourselves of Ortega y Gasset's concept of *reparar* (2004, Lesson II). *Reparar* (notice, spot, bring to attention) is the action of becoming aware of something, bringing it to the focus of attention by discerning and establishing it as distinct from the rest. *Reparar* is to create a distinct foreground against an all-pervading background. Ortega's notion of foreground (*primer plano*) becomes key for our analysis of a situation because, at any given moment in time and point in space, there is always much more happening than we can acknowledge at once. With my attention (*mi reparar*), I give a relatively stable and definite shape to a small set of relations that become highlighted over the extensive background.

Let us introduce an example à la Ortega to further explain this point. Imagine that you are a student sitting in my seminar and I am facing you as I speak in the classroom. I suddenly raise my right hand with a water bottle in it, asking: "What do you see?" You surely answer: "A water bottle." Nevertheless, the water bottle is just one among many of the things that are currently visible to you in the classroom. It is only your attention that brings the water bottle to the foreground, turning it into the protagonist of all the events, processes, phenomena, and actions that are happening in conjunction with this place and time (and which in turn constitute the perceived place and time).

In fact, while you assert that you see a water bottle, focusing all your attention on this one object I have consciously conditioned you to privilege, there is a moth stamping against one of the lamp lights, a door that slightly vibrates because of drilling done in the hallway behind it, a student scratching his head, another one hiding a yawn, a piece of paper flying into the air from the teacher's desk, lots of scribbles and arrows on the whiteboard, and a warm light coming through the side window and falling onto the floor. None of these events, entities, and processes are worthy of belonging to the field of the visible and noticeable in answer to my question, "What do you see?" You only claim to see the water bottle.[4]

Why? Clearly because I manipulated your attention by raising my hand with an object in it and making this action coincide with my question. This trick serves to demonstrate that *reparar* (focus attention on) makes being seen and, by virtue of making being seen, it also makes being. This attention is the necessary subjective element, without which a situation cannot exist. To wit, we cannot simply walk into a situation, for the situation does not exist prior to our noticing it.

The Ontology of a Situation

What, then, exactly constitutes a situation? What is its ontological status? We have already hinted at the fact that a situation is something that a subject creates with her attention. Ortega observes that attending to something leads to realizing or fully grasping (*per-catarse*) that something. In the example above, the classroom, the door, the chairs, the desk, the other students, the light, and the physical presence of the professor belonged to the field of what you passively knew was there and you accounted of—unconsciously relied on it as a background certitude requiring no attention (*contar con*) (2004, Lesson III).

Such background is formed by everything that appears without our noticing it, without demanding the slightest touch of our attention. But everything in the background, Ortega advances, has the potential to become a temporary foreground protagonist: we can always transform our *contar con* into a *reparar* by putting our attention to work.[5] This shift from a

[4] See Brook Ziporyn's discussion of "the Gestaltist premise that when some 'one' appears as an explicit coherence in the above sense, it appears as a figure against a background" (2004, 46) and Hershock's "horizons of relevance" (2004, 63).

[5] Notice that Ortega does not directly discuss the ontology of situations. He is interested in what it means to be alive, describing the task of the human as a task of radical orientation. What Ortega asks his students to bring to the field of *reparar*

mere counting-on to a full realization of certain relations as appearing is how a situation is born.

We are claiming, then, that a situation belongs to the intentional look of the subjective agent. However, any intentional look also brings to the foreground entities other than the agent herself. *Reparar* is an introspective look upon oneself (what do *I* see?) that includes entities that are conceptualized as not being the self (I see a water bottle). In noticing the world we discover ourselves, and in reflecting upon ourselves we discover the world. With Ortega, we affirm that by the time we acknowledge something (ourselves or other), the world is already out there as a background that enables and affords any of the possible experiences and situations we may raise out of it.

The given world does not need my attention to exist, yet it is inextricably connected to me. In my switching from *contar con* to *reparar*, endless possibilities of situation-making are afforded. Which means that things *are*, but they are not this nor that. They are what Ortega calls *un problema*: "The radical and irremediable fact is that living man finds that neither things nor himself have being; that he has no choice but to do something to live, to decide his doing at every instant or, what is the same, to decide his being, and this includes, as we have seen, the being of things."[6]

Situations are not what they are simply by virtue of the pre-existing elements that come to define them—agencies, entities, events, processes emerging along with a particular space-time—but they are by virtue of what they are decided to be: what a consciousness makes out of them. In this manner, the ontological status of a situation is that of not being *per se*. It has no ontology of itself.[7]

The *Zhuangzi* is a valuable source of utterances embodying this ontological claim, namely that the same "ingredients," that is, a highlighted net of relations including spaciotemporal ones, may be constructed into dis-

---

from the field of *contar con* is their own awareness of themselves (*yo*) and the world surrounding them (*mi circunstancia*).

[6] "El hecho radical e irremediable es que el hombre viviendo se encuentra con que ni las cosas ni él tienen ser; con que no tiene más remedio que hacer algo para vivir, que decidir su hacer en cada instante, o lo que es igual, que decidir su ser, y esto incluye, como hemos visto, el ser de las cosas" (2004, 224). Per convention in his times, Ortega repeatedly uses "man" and the male pronoun to refer to human.

[7] This is a claim that could be made about any given entity, the only ontology being that of the totality of interconnections and interdependencies seen as one and susceptible to become an infinite array of relational configurations by means of boundary-making.

similar, even opposite kinds of situations from an evaluative point of view, hence a situation *is* not *per se*.

One of the best-discussed stories embodying this claim features Confucius with some of his disciples being forcefully restrained and besieged between the states of Chen and Cai. The same elements (ingredients, relations) are acknowledged and brought to the foreground: there is no cooked food to eat, they find themselves in a physical state of great exhaustion, they are held prisoner, and Confucius is threatened to be killed with impunity.

However, where the disciples evaluate these elements as raising a situation of great distress or failure (*rucizhe kewei qiong yi* 如此者可谓窮矣), Confucius only sees success and good fortune (*tong* 通/*xing* 幸), the rationale being that they afford him opportunities to cultivate his moral capacity and externally perform his virtue.[8]

A second claim is that, ontologically speaking, a situation both is and is not. It is because it exists by virtue of the focus of our attention (*reparar*): what we notice and how we interpret it (given that it is *not per se*). It is also not because there is nothing to keep it together as a situation except for this momentary focus of attention. Which is to say, a situation has no essence nor identity, just temporarily imposed boundaries created out of cherry-picked relations that are brought to the fore.

These are zoomed in, and highlighted, by means of obscuring their extended nets of relations, as if we were to use a flashlight to illuminate a circle on a paper and then endow the illuminated circle with self-identity, essence, and independent existence. Everything that remains in the dark, the rest of the piece of paper but also the table where the paper lies, the

---

[8] See the anecdote in *Zhuangzi* 28 (Guo 2004, 28: 981-83). It appears in many different versions in the *Zhuangzi* and other early texts, each framed differently to illustrate a different teaching (Makeham 1998). I point the reader to the version where we can see that a situation is not determined by its elements but by how it is decided by a subjective agent.

There are many other passages that embody this claim, as the famous discussion between Zhuangzi and Hui Shi over the use of a gigantic gourd in chapter 1, the dead dialogues in chapter 6, and the deformed Shu in chapter 4. All these (and many other) passages have in common that the same relations hold different meanings and are qualified as opposite kinds of situations in good-bad/right-wrong binary systems. The *Zhuangzi* uses this anecdotes and dialogues to show that the binary itself, as any other system of classification, is a human projection and delusion. Things are not *per se*, which does not mean that they do not exist but that they do not have fixed meanings, essences, nor functions until they become determined.

floor that holds the table, my hand holding the flashlight, etc., is not the situation *per se*, and yet it is a constituting part of what we call the situation, i.e., the circle illuminated by light.

The situation, then, is and is not in a nondual manner: being by virtue of not being, like the illuminated circle is by virtue of isolating it from and not attending to everything that is not the circle and yet constitutes it. Brook Ziporyn encapsulates this idea in the expressions "to be present as X is also to be present as not-X," "there is more to any X than is known at any time," and "there is an unseen back to anything" (2004, 62).[9]

When discussing nonduality with my students, I show them the optical illusions of the duck/rabbit and the old/young lady (see Ziporyn 2004, 159). I explain that they can only see one figure at a time (either duck or rabbit, young lady or old), and yet both figures not only coexist but, emphatically, cannot be without the other. The rabbit is constituted out of duck, and the duck out of rabbit. The rabbit appears by virtue of our attention that isolates it as much as by virtue of our *desatender* (neglect, omit) that negates the duck turning it invisible.

This implies that the rabbit is so much by virtue of what it is (a rabbit) as by virtue of what it is not (a duck). The same happens with the constitution of a situation. The identity of a situation depends on what is brought to discretion by our attention (*reparar*) against the background of everything else that constitutes the situation and yet is conceptualized in a negative way as what the situation is not.

[9] As Brook Ziporyn points out, the other that is the self is what Maurice Merleau-Ponty (1908-1961) called "the invisibility *of* every visible" (2004, 63).

Only by negating the duck can we see the rabbit. Similarly, only by creating a temporarily irrelevant and diffused background can we raise a situation to the foreground. The situation is created by means of active exclusion or, in the *Zhuangzi*'s terminology, "active oblivion" (*wang* 忘). When we identify a situation, we exclude the potentiality of all that is not selected as focus of our attention and neglect all other possible forms of relations that the particular space-time in which we are inserted affords us. And yet, the potentiality of the neglected relations is necessary for what appears as a determinate and identified situation, and it is constitutive of the situation as much as our look that neglects it—counting on it but not attending to it. This actively forgotten background that enables and affords for discrete entities to appear and for situations to be discerned consists of an endless net of relations with no fixed boundaries that connect every single thing with potentially any other single thing.[10]

To use yet a different image we may think of a doodle of numerous lines that intersect, each intersecting point being an entity, constituted of relations as its primary ontology. I too am one of these points at the intersection of many relations. My asymmetrical power, which I share with other human and nonhuman animals, is that I can pick out some of these intersections and reify them, so they temporarily appear as a thing that externally and independently lies in front of me. Yet at the same time I obscure with my active oblivion all other intersections and the larger nets of relations in which these are inscribed and by which they are constituted.

This temporary reification, by means of which I create a situation, this illumination that makes visible, is interdependent with all which remains in the dark as background. That is, all that is not the situation is also relevant for the situation, and in this sense, it in fact is the situation while being conceptualized as being not. A way of describing this ontology of "intrinsic and constitutive relationality" (in Roger Ames' words) is the Chinese doctrine of *yiduo bufen* 一多不分, one is many and many is one: "It is,

---

[10] In his recent monograph (2021), Roger Ames contrasts classical Greek "one-behind-the-many" ontology, which takes *eidos* as a principle of individuation, with Chinese cosmology, which begins from the primacy of vital relationality and where everything is relevant to everything else (ch. 4). He talks of cosmology and becomings instead of ontology for the Chinese case, because he takes the word ontology to imply an essentialist view of individual entities as prior to relations. I am instead using the word ontology to simply mean "a discourse or understanding on how things are constituted." I rescue the word ontology but eliminate the Greek foundational assumption of "on-behind-the-many" to apply it to Chinese discourses on what and how things are, much like we can rescue the terms agency, ethics, or subject and redefine them, seeing them evolve in usage and meaning.

simply put, the assumption that in the compositing of any 'one,' there is implicated within it the contextualizing 'many'" (2021, 218). The principle of individuation is misleading: where we see one combination of relations raising as one situation, there are in fact numerous available combinations of relations at work, or again in Ames' words, every focus has (and is constituted by) a field.

Boundaries

This raises the question of the possibility of a situation being bounded, of having boundaries. If a situation (via focus, attention) is partially by virtue of what is not (via field, background), and the relational field of what is not is limitless and unbounded, then how can a situation come to be limited, determined, and brought to presence by means of individuation?

We may borrow from Gilles Deleuze (1925-1995) and Félix Guattari (1939-1992) the metaphor of a node in a field of forces or rhizome to describe how something appears as individuated while being a product of the collective: constituted and constitutive, in equal parts, of the totality (Deleuze and Guattari, 1980). How do these nodes in a field of entangled forces, or maybe knots in a net, or even hot spots in an ever-expanding continuum appear? Let us explore the arising of nodes, knots, or hot spots that individuate a situation as a thing happening out there against an actively forgotten background as concentrations of coalescing relations.

Once again relying on Ortega, we may indulge in an ordinary-life example. You are riding a bus back home from an excursion to the new park in the city. You are listening to music and vaguely looking through the window. Nothing is happening, except that so much is being given: a teen is reading a newspaper, another is looking at his phone, the bus driver keeps wiping sweat off her forehead with her sleeve, it smells like dirt and oil, you are digesting an apple, the chair in front of you looks worn out and discolored. And yet, nothing is happening for you: nothing is condensed enough to constitute a situation. There is no sufficient concentration in any of the described space-time relations and affordances to be brought to the fore, to demand the passage from *contar con* to *reparar*.

Notice that these relations would be more than enough to constitute an unforgettable, distinct, and well-bounded situation for a two-year old riding the bus for the first or second time—but not for you. Notice also that all these relations, including their particular intersection with you, are more than enough to constitute an unforgettable, distinct, and well-bounded situation for Karen who has never seen a black transgender person riding her bus before—but not for you.

Then you get a phone call, your friend asking, "What are you up to?" You reply, "I'm on the bus riding home." You just created the situation "on the bus riding home" by privileging certain space-time relations in order to bring yourself to presence along with what you deem to be your most relevant circumstance with regard to your friend's inquiry. In this case, your friend's question is the trigger to cause a condensation of certain relations that become privileged against a background turned irrelevant.

As you hang up the phone, the bus comes to a sudden and violent stop. Everyone gasps and shakes in fear and disconcert, grasping handles and seats. You look through the nearest window and see a bleeding person on the ground, probably hit by the bus or by a green car stopped beside it. You get off the bus along with the other passengers, grab the phone, and call back your friend: "There's been an accident."

Meanwhile, the sky is blue and radiant, sun rays heat up your skin, a group of sparrows of different sizes pick crumbs from a table, a little girl rushes down the sidewalk on her scooter wearing a helmet, classical music sounds off a balcony, the new sandals you are wearing make your feet hurt. The condensation of the relations that become "an accident," however, is such that it saturates your attention and creates an irresistible and all-excluding knot, hot spot, or node. Even the situation of "riding the bus back home" has now become just a germ, a history, a root, or a context for the true situation that "there's been an accident."

The knot appears because of our *reparar*, but this *reparar* may have different causes: self-directed awareness (I am to observe my breath for the next ten minutes), other-provoked awareness (What are you up to? What do you see?), or impromptu happenings that condensate or even saturate our attention because of their surprising, dangerous, or demanding nature (from a burgeoning fire that must be immediately extinguished to my daughter repeatedly and loudly requesting her snack). In this way we vindicate Ortega's claim that awareness and *reparar* are caused from encounters with *problemas*: events and phenomena that present resistance and imperatively demand our resolution.

This is also what Charles Sanders Peirce (1839-1914) called "the irritation of doubt" (1877, IV). It is the itchy feeling that something needs resolution and forces us to struggle until it is resolved. He discusses doubt as an unpleasantness that leads us to do anything to escape that state by reaching a belief, possibly defined as a condensation of relations associat ed with a fixed meaning. This being so, the selection of matters for attention that become individuated as situations is a response to "problems" to be resolved. This holds true as long as we understand *problema* in the Ortegan sense of what irritates us, because it yet needs to be decided and defined—thus he characterizes life itself as an open project.

We need one extra qualification on how a condensation, node, knot, or hot spot of relations appears to raise a situation via our *reparar*. As hinted above, riding the bus was nothing for you, but it was a distinct and memorable situation for the toddler and the white lady (as long as you were on the bus), much as the accident provoked a saturation of attention for you and the lady yet not for the toddler.

The relational affordances that we actualize and reify at any given time as constituting a situation depend not only on what we are forced to see by things' resistance or imperative for resolution (their being a problem), but also by what we are trained, educated, and socialized to see—and not to see—when we look at and around ourselves.

Much like Karen at the sight of a black transgender person on her bus, Confucius freaked out at the sight of a someone leisurely and joyously swimming in the massive waterfall of Lüliang. He could not conceive of this occurring as anything other than an attempt at self-harm, unaccustomed as the great master was to demonstrations of natural adaptation to one's environment or acquired, non-learned virtue (Guo 2004, 19:565-58).[11] Our expectations and conceptions of the normal and the good determine what catches our attention and also how we are to interpret those condensed relations that obscure everything around them.

Co-Constituting

We are to consider with the *Zhuangzi* that our current set of expectations, beliefs, and values is the result of previous encounters and events that help "fully form our heart-minds" (*chengxin* 成心), namely making up our minds on rights and wrongs (*shifei* 是非), possibles and impossibles (*kebuke* 可不可) (Guo 2004, 2:56). The I who creates and raises a situation is in turn itself a product of a series of previous situations which dictate the kind of situations it will in the future co-create. Situations and I, I and situations co-construct one another over time. The *Zhuangzi* says, "Without the other there is no I, but without I there is nothing to grasp" (*feibi wuwo, feiwo suoqu* 非彼無我，非我無所取; Guo 2004, 2:55).

Let us explore the first part of 0
claim, which I believe to be more difficult to accept than the second. Indeed, after almost a century of social constructivism the European history of transcendental idealism, or perhaps simply because we have access to our consciousness, we can have a first-person experience of how our *reparar* and *atender* makes things be and become something and, as a result, we do not have such a hard time understanding just how we create things ("without I there is nothing to grasp").

But how is the other ("perceived and conceptualized as different from I"), in this case a situation, constitutive of the I ("without the other there is no I")? As Roger Ames explains, not only are fields constituted by their foci—such as history by events, families by their members, and, we may now add, "situations by persons"—but foci are also constituted by their fields, that is, events by history, members by their families, and persons by situations (2021, ch. 4). So, how am I made out of things that are not I?

There are at least two relevant senses in which situations constitute the person. The first comes from a narrative conception of the person and can be summarized as the process of acquisition of a fully-formed heart-

[11] In the story, Confucius qualifies this situation as "suicide attempt" and sends his disciples to help, not being capable to even conceive that the swimmer is not in need of help. He is consequently ridiculed by the adept swimmer who, without giving it much importance, claims to know how to swim in those waters simply by living in them. See Galvany 2019.

mind. This involves the creation of horizons of expectations and conceptions of normality, which inevitably are associated with axiological evaluations of good and bad, right and wrong, possible and impossible. What I have experienced in the past influences and acts upon what I will experience in the future and how I conceptualize it and endow it with meaning.

My horizon of expectations dictates whether something is happening, whether something is to be seen, thought, considered, included, learned, or known. In other words, my previous situations—all which I am not, such as "there's been an accident"—dictate who I become and the kinds of situations I will co-raise in the future.[12]

We are the product of our histories, both actors and patients in them. As Roger Ames remarks alluding to Alfred North Whitehead's (1861-1947) holistic aesthetic order,

> It begins from the assumption that all of the concrete and interpenetrating details of this particular painting and its unbounded context are relevant to the totality of the effect. When we move from paintings to persons, we must acknowledge that all of the narrative details—the entire field of events of our lives—are more or less relevant to the emerging identities of whom we are becoming as persons. (2021, 211)

The second sense comes from a transformational perspective or "transformation of entities" (*wuhua* 物化), the notion that every entity, defined as an emerging collection of interconnected relations with varying degrees of interdependency rather than as an individual substance, experiences continuous transformation along with changes in its constitutive relations. In his analysis of the *Zhuangzi*'s butterfly dream (Guo 2004, 2:112), where the term *wuhua* appears, Dan Lusthaus claims that different situations radically transform the subject of experience that raises and constitutes those situations (2003).

The dream makes us witnesses of such transformation where an entity called "I"—an intersection between plural and changing relations—experiences transformation between becoming a human subject named Zhuang Zhou and a butterfly subject. Lusthaus explains that each situa-

[12] Unless I engage in an active self-cultivational work of self-aware deconstruction which may counteract the effects of my socialization, as the *Zhuangzi* describes with the process of "sitting in oblivion" (*zuowang* 坐忘; Guo 2004, 6:282-85). It is very much aware of the absolutist dangers of socialization but also acknowledges that there is no living without socialization and mediation. The proposal is to deconstruct the fixation of the items that have become reified as to return to them (and to oneself) their original ambiguous potentiality of being nothing per se.

tion carries its own distinct set of rules or, in the *Zhuangzi*'s words, "divisions" (*fen* 分), as they become individuated and singled out from a background.

It is obviously not the same to act as a person or as a butterfly, but it is neither the same to act as a writer—as I am doing right now, much aware of my situation as I actively self-direct myself to reflect on it—or to act as a mother—a role I am constantly forced into as I write with my young daughter at home. The subject, which creates a situation by noticing and bringing certain relations to the forefront, is in turn created along the situation that gains primacy and rises. Despite her asymmetrical power to reify relations via attention, the subject or I is just one more constitutive element in a situation, as dependent upon the rest of elements and relations as these depend upon her *reparar* to become.

Hence the person changes as much as all relations change and along with them, although this transformation is not always so radically visible as in the boundaries between a person and a butterfly. As Lusthaus notes, "Transformation involves radical novelty, such that it is not that a self-same object goes from situation A to situation B, but that person A in situation A becomes something else (butterfly, natural phenomenon, etc.) in situation B" (2003, 170).[13]

Normative Considerations

Why is it important to understand the ontology of a situation? To my mind, theoretical reflection is interesting and engaging as such, a revealing exercise that helps us see less or more than what we ordinarily see. What makes it relevant and necessary, however, is its direct impact on doing, interacting, performing, and behaving. Different ontologies lead us to different models of what it means to act and live well (with)in this world. It is because of its guiding and normative power that I consider ontology and other branches of theoretical speculation more than an entertaining and eye-opening exercise: they substantiate and legitimize the kind of persons we are to be and how we are to treat others, and therefore must be taken seriously.

---

[13] A critic may say that the person remains the same and only the action, role, or attitude changes, but that would be an essentialist approach to persons and entities as antecedent substance prior to relations for which we find no evidence to sustain. A change in situation represents a change at the ontological level of the constitution of the subject/agent/person/I as well as the rest of its constitutive interconnected relations.

By the time we ask a question about life we are already in it, said Ortega, which means that all our questions occur *a posteriori*. We find ourselves betwixt and between, always late to our appointment with the world. By the time we can inquire about who and how we are, we already are. By the time we come up with an answer about who we are, we have probably become something else.

At stake here is an ever-present primordial entanglement that cannot be unknotted. Many forms of philosophical analysis attempt to isolate discrete elements in complex relations in order to understand them separately and independently under the assumption that reduction to individual parts simplifies the task of thinking and leads us to the nature of things. I am starting from the opposite assumption: things are messy, intertwined, embedded, entangled, interconnected, interdependent.

There are not even "things" in the substance-ontology sense of the term, but only relations with hot spots that enjoy differing degrees of stability. We need theories that help us understand who and how we are in this messy, entangled, interconnected, relational manner, so that this understanding can help us devise more efficacious ways of thinking, feeling, and acting that are based upon our ontological status, on what we are. We are in the middle of intricate nets of relations that gain and lose temperature and condensation, both by means of these relations and constituted by these relations.

We are in and by the entanglement, an entanglement so intricate and vast that, as the *Zhuangzi* says, we can never know where anything begins nor ends, what constitutes being or presence and what constitutes nothingness or absence (Guo 2004, 2: 79). The moment we identify a beginning, a new beginning for that first beginning can be identified just by enlarging our perspective on its constitutive field, an exercise that can be repeated *ad infinitum*.[14] What our focus of awareness now deems a being quickly turns into nothing when we shift our attention to a brighter and newer stimulus, leaving our former being into the actively forgotten background of what eludes our *reparar*.

Beginning and end, being and nothing are both two and the same; they turn into one another just like the rabbit and the duck by the simple switching of our attention. A situation is just a focus of awareness upon certain relations which can always be ever extended by illuminating a larger focus or reduced by concentrating on ever-smaller relational fields.

---

[14] This insight is crucial for Kyoto School founder Nishida Kitarō's concept of place (*basho* 場所). See Nishida 2012.

If this is so, nothing is never just what it appears to us at first sight nor is it ultimately the meaning we ascribe to it.

The *Huainanzi* 淮南子 parable of the border man who lost his horse makes us realize that we never know whether what we notice to be happening (a situation) is fortunate or unfortunate, a beginning or the end of something, a thing or nothing (Liu 2003, 14:597-99).[15] The events play out to disprove our conventional notions of fortune, right, and good by simply adding a bit more of awareness over context and entangled relations each time.

In the story, a man's horse runs across the border into a different people's territory, that of the Hu, from where it cannot be recovered. Everyone (*renjie* 人皆) pities his loss, but he responds, "How do you know this doesn't constitute fortune?" Several months later, the horse returns bringing an excellent Hu steed along, and everyone rejoices. The man, however, asks, "How do you know this doesn't constitute misfortune?"

Next, his son falls while riding and breaks his thigh, to which everyone reacts in dismay. Once again, the man wonders, "How do you know this doesn't constitute good fortune?" A year later, the Hu invade the area and all able-bodied men are drafted. Nine out of ten die, but not the man's son, who could not serve due to his lame leg.

The story shows how fortune and misfortune are interdependent, nondual just like the rabbit and the duck—each constituted out of the same set of relations and available to be brought to the fore depending on our *reparar*. It contrasts conventional morality and standard axiological evaluations by teaming everyone against "the man," who is described as an expert in mantic arts, implying that he understood that there is always more beyond the small frame of what appears to be in the here and now. He knows that things are neither this nor that, but an open problem without any essence of their own. They only become fixated and determined via our focusing contextualization. In consequence, moral and critical evaluations must vary, and he maintains an open attitude, refraining from short-sighted extreme reactions like those of "everyone."

Things are messy, intertwined, interconnected, entangled relations with no fixed beginnings nor ends. However, we can and do privilege starting points every time that we deem something so, affirm something right, or notice that something happens—we raise a situation. That is how we create worlds (*mundos*): the privileging of a knot to act as a starting

[15] This *chengyu* 成語 (phrase, idiom) inspired by the *Huainanzi* story— *saiweng shima* 塞翁失馬, "a man on the border loses a horse"—remains in Mandarin today, meaning that a loss may turn out to be a gain, or vice versa.

point, as a source of narrative and causal meaning for a larger set of relations.

Ortega says, "World is that of which we are certain."[16] What a magnificent thought! We cannot live in the messiness of our circumstances. Without a starting point, a fixed anchor, a set of beliefs and values, a standard against which to measure things. Reality is like moving water—we fall, we sink, we are drawn down. Insecurity (Peirce's "irritation of doubt") forces us to fabricate security: certainty, a solid stepping stone. Ortega insists that it is our own believing in our own fabrication that saves us from drowning, that keeps us safe. This is why against an essentially inhabitable reality we establish a world that is habitable by virtue of our certainty, of the trust we put into it. When we build safety nets against the itching unpleasantness of incertitude, we create worlds that come to possess causal power by being shared by many: they all intersect yet never fully overlap.

But what happens when we are so certain of something? When we trust something to be so, true, and right? We become blinded by our own comfort and unable to see, affirm, and find existential security in alternative yet equally legitimate worlds. The *Zhuangzi* equates the worlds we raise through certainty, our stepping grounds for the creation of meaning, with points in a circle.

Each point is a beginning, from which to interpret with security our lived experience. Each point has grounds on which to be formulated, legitimized, affirmed, and accepted, but by the same logic each point also has grounds on which to be falsified, negated, disproven, and rejected (Guo 2004, 2:66). Each point is a position that leads to affirming certain situations and endowing them with particular meanings via our belief—defined as a condensation of relations associated with a fixed meaning.

This entails that each point and the situations and axiological evaluations it raises have their own enabling and limiting factors. They make certain standards emerge or submerge, be visible or invisible, possible or impossible, right or wrong, under particular conditions. As the *Zhuangzi*'s says:

> Among things, there is none that cannot be seen from "that" position, and none that cannot also be seen from "this" position. From "that" position, ["this" position] cannot be seen. Depending from which position you approach something, you will know an aspect or another of it. Therefore, it is said: "that" position comes from "this" position, and "this" position also ex-

[16] "Mundo es aquello de que estamos seguros" (2004, Lesson X).

> ists because of "that" position. [The existence of] "this" and "that" is what we call co-dependent origination. Although that is the case, as things live they die, and as they die, they come to life again; things that are possible are also impossible, and being impossible, possible they become; having reasons to affirm is having reasons to deny, and those reasons to deny mean that there are reasons to affirm. (Guo 2004, 2:66)[17]

Each one of the many possible worlds is but a tiny point in the circle that, by its own raising, closes up and obscures the endless number of alternative positions from which to look at ourselves, feel an emotion, think a thought, react toward an event, interact with another. Each point is a position of openness and closure at the same time.

By enabling my seeing something, it blinds me to all that, from that position, cannot be seen. I claim to see the water bottle in detriment to the moth, the vibrating door, the ray of light, my own body. As Dan Lusthaus exclaims, "What limits us is ironically the very absence of limits!" (2003, 185). Now a butterfly now a Zhuang Zhou, now a rabbit now a duck, now "this" now "that," but never both at the same time.

We need to focus, determine, choose, and take positions. We can make worlds thanks to these starting points we secure and occupy, but we are equally blinded by them, since they force us into obscuring all other possibilities that cannot be simultaneously acknowledged for my world to make sense and function well. The normative aspect consistent with the ontology of a situation is that there is always a plurality and heterogeneity of values, standards, worldviews, and worlds of experience. They coexist and are constantly available to us, but we rarely take advantage of them.[18]

Most people do not even take joy in knowing that these plural and heterogeneous worlds exist and, as a result, work hard to deny their legitimacy and see them enclosed behind walls to prevent them from overflowing. The great María Lugones (1944-2020) repeatedly denounced this phenomenon in her work, while also offering practical advice on how to travel between worlds with a playful attitude (1987).[19]

---

[17] 物無非彼, 物無非是. 自彼則不見, 自知則知之. 故曰：彼出於是, 是亦因彼. 彼是, 方生之說也. 雖然, 方生方死, 方死方生；方可方不可, 方不可方可；因是因非, 因非因是.

[18] Dan Lusthaus discusses this important point while analyzing chapter 17 of the *Zhuangzi* (2003, 173).

[19] I heard the sad news of her death as I was completing this article in July 2020 and could not resist the opportunity to pay homage to her, however short and simple.

Living Temporally

As much as we need our world to be safe and certain in order for it to be habitable, there is no point in fixating our standards and values onto something that is contingent and partial, namely onto something that by definition is not and will keep changing along changes in its own constituting relations, eventually disappearing and never returning to presence in the exact same shape (like youth, like health).

There is no point in affirming something as absolute and true just because it is the easiest straight peek from our window. There is no point in imposing our lived experience and situations over all other lived experiences and situations that are created along with similar space-times. There is no point in spending our life mourning for what we are not, cannot have, is gone, or on the other side of the line of time, anticipating what is to come, we will become, and hope to achieve. Of course, there is no escape (nor need to escape) from noticing and bringing to present through our attention, from creating entire worlds made up of situations where everything is fixed, monolithic, uni-dimensional, and straightforward, where what you see is what you get.

Like the *Zhuangzi* says, if we follow our fully-formed minds and make them our teacher, who could ever be without a teacher (read "authoritative guide") (Guo 2004, 2:56)? Both the intelligent and the fool create fixated worlds just by selectively foregrounding relations out of the totality. We cannot live in nothingness: we would sink and drown, so we make ourselves masters of our worlds. Each point, each moment, is filled with a determinacy: my this, my right, my so. But "to claim that there are any such things as right and wrong before they come to be fully formed in someone's mind, that is like saying you left for Yue today and arrived yesterday," that is, self-contradiction, nonsense, or a play in words (Guo 2004, 2:56; Ziporyn 2009, 11).[20]

The person who refuses to inhabit just one point in the circle but relocates to the center (*huanzhong* 環中), from which each point—position, perspective, worldview, set of standards—is equally accessible and easy to let go of, does not commit to any single world. Such a person refuses to fix the meaning of right, good, and possible, and refrains from reifying situations and endowing them with a closed meaning.

If we keep in mind the empty ontological status of the situations that define what we are (not being per se, and being by virtue of not being), we will find it easier not to absolutize our reduced, biased, and contingent

[20] 未成乎心而有是非, 是今日適越而昔至也.

views, and to switch from one view to another as needed. The *Zhuangzi* puts it most appropriately:

> Therefore, the sage does not proceed from this [vicious circle of co-dependence], but gets illumination from heaven so that his affirming "this" is adaptive. His this is now a that, and his that is then a this. His that includes something to affirm and something to deny, and his this also includes something to affirm and to deny.
>
> So, in fact, does he still have a that and this? Or does he not have a that and this anymore? When this and that do not find themselves as opposite positions, this is called the axis of Dao. The axis obtains the position of the center of the circle, and uses it to respond without limits. His affirming also responds without limits, and his denying also responds without limits. Therefore, it is said: "There is nothing like using clarity." (Guo 2004, 2:66)[21]

Ontology gives us clarity. We can never illuminate the totality of the constituting field of relations for any given focus. In this way, the position at the center is not an absolute one in terms of knowing and seeing—as the term "sage" may imply for the reader. Rather, it is a methodical position that guides how to react to our own views, ideas, and feelings as they arise and how to create paths for relativizing and enlarging our understanding of them. This in turn informs us how to act toward any given situation and any given other, and how to move in between worlds.[22] We humans have an incredible capacity to create situations, not only to deal with them or to walk into them, but to create them out of the endless indeterminacies that are available to us at any given space-time—always at hand (*a la mano*), always possible.

---

[21] 是以聖人不由, 而照之于天, 亦因是也. 是亦彼也, 彼亦是也. 彼亦一是非, 此亦一是非. 果且有彼是乎哉？果且無彼是乎哉？彼是莫得其偶, 謂之道樞. 樞始得其環中, 以應無窮. 是亦一無窮, 非亦一無窮也. 故曰「莫若以明」.

[22] The person located at the center of the circle is just occupying a position for situation-opening and boundary-creation, and he knows it. There are limits to what can be illuminated at any single time: limits by positionality, relationality, and perspective, and even those who illuminate more cannot illuminate the entire totality of the cosmos at once. Thinking through the image of "getting illuminated by Heaven," even such major light that illuminates half of earth fails to illuminate the other half and itself (heaven). Much as with yin and yang, the sunny side depends on the shadowy one, and it would take a zero absolute perspective to see the whole at once, a premise that is not given nor accepted here.

Bibliography

Ames, Roger T. 1998. "Knowing in the *Zhuangzi*: From Here, on the Bridge, over the River Hao." In *Wandering at Ease in the Zhuangzi*, edited by Roger Ames, 219-30. Albany: State University of New York Press.

_____. 2021. *Human Becomings: Theorizing "Persons" for Confucian Role Ethics*. Albany: State University of New York Press.

Barad, Karen. 2006. *Meeting the Universe Halfway. Quantum Physics and the Entanglement of Matter and Meaning*. Durham: Duke University Press.

Deleuze, Gilles, and Félix Guattari. 1980. *Mille Plateaux. Capitalisme et Schizophrénie* 2. Paris: Les Éditions de Minuit.

Galvany, Albert. 2019. "Panic, Parody, and Pedagogy at the Waterfall. Morality as a Misleading Principle for Moral Action." In *Skill Mastery and Performance in the Zhuangzi*, edited by K. Lai and W. W. Chiu, 201-27. London: Rowman & Littlefield.

Guo Qingfan 郭慶藩. 2004. *Zhuangzi jishi* 莊子集釋. Beijing: Zhonghua shuju.

Jullien, François. 1995. *The Propensity of Things: Toward a History of Efficacy in China*. New York: Zone Books.

Liu Wendian 劉文典. 2003. *Huainan honglie jijie* 淮南鴻烈集解. Taipei: Wenshizhe.

Lugones, María. 1987. "Playfulness, 'World'-Traveling, and Loving Perception." *Hypathia* 2.2:3-19.

Lusthaus, Dan. 2003. "Aporetics Ethics in the *Zhuangzi*." In *Hiding the World in the World: Uneven Discourses on the Zhuangzi*, edited by Scott Cook, 163-206. Albany: State University of New York Press.

Makeham, John. 1998. "Between Chen and Cai: Zhuangzi and the *Analects*." In *Wandering at Ease in the Zhuangzi*, edited by Roger Ames, 75-100. Albany: State University of New York Press.

Marion, Jean-Luc. 2001. *De Surcroît: Études sur les phénomènes saturés*. Paris: Presses Universitaires de France.

_____. 2002. *In Excess: Studies of Saturated Phenomena*. New York: Fordham University Press.

Moeller, Hans-Georg. 2006. *The Philosophy of the Daodejing*. New York: Columbia University Press.

Nishida, Kitarō. 2012. *Place and Dialectics. Two Essays by Nishida Kitarō*. Translated by John W. M. Krummel and Shigenori Nagatomo. New York: Oxford University Press.

Ortega y Gasset, José. 2004. *¿Qué es Filosofía? Unas Lecciones de Metafísica*. México City: Editorial Porrúa.

Peirce, Charles Sanders. 1877. "The Fixation of Belief." *Popular Science Monthly* 12:1-15. Accessed online at: http://www.peirce.org/writings/p107.html

Searle, John R. 1995. *The Construction of Social Reality*. New York: Free Press.

Ziporyn, Brook. 2004. *Being and Ambiguity. Philosophical Experiments with Tiantai Buddhism*. La Salle, Ill.: Open Court Press.

_____. 2009. *Zhuangzi: The Essential Writings with Selections from Traditional Commentaries*. Indianapolis: Hackett Publishing.

Time and Space

within Daoism's Holistic Worldview

JOSEPH L. PRATT

The *Daode jing* offers a complete account of reality, running from an ultimate emptiness, which is called Dao, to the multitude of everyday things. Central to the Daoist account is the yin-yang dynamic, which can exist in a state of either harmony, in which case Dao is realized, or discord, in which case life is defined by a series of hardships. The seamless explanation encompasses consciousness, akin to Dao, as well as cognition and form, both necessary for the experience of Dao. This elucidation also intertwines time and space, which too have yin and yang characteristics and feature in harmonious and discordant states. The yin aspect of time and space manifests as the deep here and now, while the yang aspect appears as the conventional past and future or near and far.

When the central yin-yang dynamic is in harmony, consciousness, cognition, and form converge in a state of effortless action (*wuwei* 無為). In this state of oneness, the here and now is realized and linear time and space are experienced but for this deeper truth. In discord, on the other hand, consciousness, cognition, and form struggle among each other so that a person suffers unease and life becomes difficult. Time and space are experienced as a conflict between the need to be present and the need or desire to be some other place and time. Nearly all sense of the abiding here and now is lost. The ideal state of harmony with time and space dilation has been well-documented and is characterized by modern positive psychology as the state of "flow." The disruptive state of discord, on the other hand, is now often misconstrued as the basic human condition.

The first part of this chapter lays out the *Daode jing's* holistic account of reality. Following the oft-cited chapter 42, it explains how reality unfolds from ultimate emptiness first to a transcendent oneness, then to the central yin-yang, and finally to everyday form. This seamless account is the opposite of the modern reductionist approach to reality which can be traced back to Sir Isaac Newton (1642-1726) and his misconception of time and space as absolute.

The chapter next places consciousness, cognition, and form within the holistic context. These paragraphs show how consciousness manifests as first cognition and then form, and these aspects can exist in a state of integration allowing for effortless action or a state of disintegration in which case life is experienced as hardship. In connecting these three aspects, it shows how Daoism solves the mind-body dilemma associated with Rene Descartes and explicates the hard problem of consciousness beguiling modern artificial intelligence research.

The third part of the chapter turns to time and space within the holistic account. Time and space too are a function of consciousness and then cognition and form. In explaining how time and space involve both the absolute here and now and the relative past and future or near and far, it demonstrates why Newton was wrong to regard time and space simply as absolute but then how Albert Einstein (1897-1955) was also wrong to regard them merely as relative.

The chapter concludes by tying the Daoist account of reality, including time and space, with modern positive psychology's well-documented experience of "flow." It indicates, as Daoist texts noted long ago, that a highly conscious state with a full sense of time and space is key to success not just in extreme activities like auto racing and rescue operations, but also in ordinary matters like managing a hedge fund or, as the *Zhuangzi* conveys, butchering an ox. This part further notes how meditation practices, such as Daoist-based taiji quan and Indian yoga, raise consciousness and thereby help practitioners gain a full sense of time and space and immerse themselves in the experience of flow.

Holistic Explanation

Though historical figures within the Daoist tradition have not always considered it, Daoism's seminal texts provide a holistic explanation of reality (Pratt and Liu 2018). The *Daode jing* account begins with an ultimate reality, which cannot be described but is commonly called Dao, usually translated as "the Way." Dao is beyond all attributes and best characterized as ultimate nothingness, emptiness, or darkness. It may also be described as the Absolute with an upper-case "A" or the Truth with an upper-case "T," recognizing that both are beyond ordinary classification.

As the Daoist sages no doubt realized, a complete explanation of reality must begin with the ineffable emptiness: a consideration of what lies behind and is responsible for objects as well as time and space as we know them. In order to experience Dao as Absolute, however, this ultimate reality must give rise to that which complements it. Dao must beget what has attributes, that is, form, as well as conventional time and space.

As the text lays out, the first step in giving rise to form involves Dao begetting the One (ch. 42). In Daoism's holistic account, the One is a totality, or complete whole. In contrast to Dao as an ultimate Absolute with an upper-case "A," the One is a transcendent Relative with an upper-case "R." Whereas Dao is Truth, the One is correspondingly Artifice. Like Dao, it defies simple categorization; unlike Dao, it can at least be fathomed.

In numerical terms, the One is simply the number one, recognizing that in a holistic explanation, numbers entail both a higher qualitative aspect and a lower quantitative one—One as a totality and one as a basic unit of measurement. The *Daode jing* appreciates that Dao and this totality are both the same and different (chs. 10, 22). The first step of the One is distinct from the ultimate reality of Dao but at the same time ultimate reality is all that truly exists, so that the One by definition knows no separation. Because the totality yet knows no distinction, however, Dao and the One cannot themselves create physical form, by which Dao might be experienced.

In the second step in the emergence of form, the One begets the Two, called the yin-yang. Yin reflects the hidden Dao, while yang reflects the apparent One. At the layer just below totality, the yin-yang Two is a further absolute-relative or truth-artifice dynamic. Within a holistic metaphysics, this layer of the Two is both a unity, without separation between the two parts, and a distinction, with an ostensible separation between them (e. g., ch. 10). In this respect, the Daoist sense of opposition differs from the modern dialectical or dichotomous sense, where the two sides necessarily contradict each other (but can somehow still achieve the transcendent One) (Pratt and Zhao 2019). The key to the transcendent One is complementarity, not contradiction.

In mathematical terms, this quality is the number two, appreciating again that numbers possess both a higher qualitative value and a lower quantitative one—the Two as a monism-dualism allowing for a full return and the two as a number of units. Reflecting the primordial interaction of Dao and the One, the yin-yang Two encompasses all dichotomous relationships, such as consciousness and cognition, attention and intention, or intuition and logic. The yin-yang Two, however, is still not enough for the full emergence of form and experience of Dao.

The final step is for the yin-yang Two to give rise to the embodied Three, again as expressed in subtle numerical terms—the Three as a formation providing for the experience of Dao and the three as a number of units. The Three allows for common three-dimensional form, as in material objects, as well as three-dimensional abstractions, which could be called conceptual matters.

In physical terms, the Three is a reflection that a concrete object cannot be considered from more than three dimensions (length, height, and width) without moving to a different scale, which would then also involve three dimensions. In conceptual terms, the Three means that an abstract concept always entails three dimensions (e. g., age, height, and weight, or logos, pathos, and ethos). With the embodied form of the Three, conceptual and physical forms may arise for the experience of the yin-yang as well as the transcendent One and the ultimate Dao.

Mirroring Dao and the One, and in accordance with the yin-yang property, the yin-yang Two is itself a yin quality, necessarily entailing and working with the embodied Three as a yang quality. The yin-yang Two reflects Dao in its essential hidden characteristics and follows the One, which is a yang property. On the other hand, the Three represents the One in its essential visible characteristics and follows the yin-yang Two, a yin quality.

The yin-yang Two and the embodied Three comprise the manifest layer of reality, a final tier of the absolute-relative or truth-artifice dynamic. In the language of modern physics, the Two and the Three are energy and matter, with vital cosmic energy (*qi* 氣) as central to form (*xing* 形). The yin-yang Two allows for a return to the transcendent One and ultimate Dao, while the embodied Three constitutes the ostensible separation for that experience to be meaningful.

With the yin-yang Two and the embodied Three, or energy and matter, all forms may emerge from the granular to the galactic. As the holistic account provides, the numbers after three constitute the variety of things, both conceptually and materially. The numerical properties zero through four, with four as precipitating form, represent a yin genesis which sets up the numbers five through nine as a yang manifestation. The numerical properties five through nine in turn correspond to the numerical properties zero through four and, as the *Huangdi neijing* 黄帝内經 (The Yellow Emperor's Inner Classic; 2nd c. BCE) conveys, constitute the five phases in Chinese philosophy (ch. 4). As with the numbers zero through nine, and following the fundamental yin-yang property, all numbers have both qualitative and quantitative properties. In this way, the holistic cosmology and metaphysics allow for an everyday reality of both absolute simplicity and maximum complexity.

For a particular form to experience the totality of all life and ultimate Dao, the underlying yin-yang dynamic must be in a complementary, harmonious state (*yinyang pingheng* 阴阳平衡). This condition is usually referred to as according with Dao and could be characterized as following the Middle Way, recognizing that this phenomenon is not simply a middle line between two opposing sides (as some Confucianists would have it) but a transcendent convergence leading to Dao. As the *Huangdi neijing* explains, for the experience of the Middle Way, the yin "unity" aspect must be stable and the yang "distinction" aspect secure (*yinping yangmi* 陽平陰秘) (ch. 3). In such a state, the yin and yang aspects—whether absolute and relative or truth and artifice—complement, support, and attain each other (Wang 2012, 178), and the transcendent One and the ultimate Dao can be experienced. With this transcendent possibility, the whole is greater than the sum of the individual parts.

For this experience to be meaningful, however, the particular form must also be able to deviate from Dao. This condition is marked by a conflict, such as in the yin-yang opening and closing (*yinyang kaihe* 阴阳开合). This discordant reality is, not surprisingly, referred to as "going against Dao." With the yin "unity" aspect unstable and the yang "distinction" aspect insecure, reality is experienced as a sense of alienation and insecurity. Rather than complementing and becoming each other, the yin and yang aspects (again, whether the absolute and relative, or truth and artifice) not only contradict, but also chase and deplete each other (2012, 180).

In such a state, truth and artifice may devolve into a regressive conflict between the valid and the false (Pratt and Zhao 2019). The experience of life can become so difficult that form as in a person hits rock bottom and has no choice but to surrender the misguided sense of reality and return to Dao. Though some Western philosophers, perhaps most notably Georg W. F. Hegel (1770-1831), have posited that such contradiction can lead to the Whole, Daoism shows that contradiction is always negative-sum; instead, the key to the positive-sum return is complementarity (Pratt and Zhao 2019). Although form can deviate, however, nothing can separate entirely from Dao (*Zhuangzi*, ch. 2). As the holistic account entails, ultimately there is only Dao.

To sum up, ancient Daoist texts provide a complete account of reality, from ultimate emptiness through transcendent oneness and a central yin-yang dynamic to three-dimensional cognitive and physical formations. All of reality is essentially archetypal but also pragmatic and exists only for the experience of Dao. With the central yin-yang dynamic in balance, the form may return to complete wholeness and ultimate emptiness. With the yin-yang out of balance, on the other hand, the form experiences reality as a series of hardships. Crucial to the experience of this transcendent Middle Way is not simply form *per se*, but the consciousness and cognition that underlie such form.

Consciousness, Cognition, and Form

Daoist cosmology and metaphysics explain how a form, as in a human being, can become conscious and cognizant of its self as well as of the Middle Way and ultimately Dao. Fitting consciousness and cognition into the holistic account of reality, Dao could be characterized as an ultimate Consciousness, while the One is a transcendent Cognition or cognitive matter. At the next energetic layer, the yin-yang Two is a consciousness-cognition, with cognition again being a "cognitive matter."

At the level of actual form, the yin-yang Two is consciousness that incorporates cognition while the embodied Three is corresponding physical matter. As this holistic explanation demonstrates, consciousness transforms first into the conceptual and then the physical. The first chapter of the *Daode jing* talks about how naming, a cognitive function, is the mother of the ten-thousand things. According to the yin-yang dynamic, moreover, the physical can transform back into consciousness, just as the cognitive transforms back into consciousness. This process is known as transformation (*zhuanhua* 轉化) (Wang 2012, 11).

As this metaphysics indicates, every form, from a subatomic particle to a planetary system, possesses both consciousness and cognition. Within any particular form, consciousness and cognition also exist at multiple levels. Within human beings, for example, they appear not just at the overall level but also the molecular, cellular, and muscular levels. At the muscular level, modern researchers sometimes mischaracterize this phenomenon as "muscle memory" (Schmidt and Lee 2005). Just as the various forms within a body constitute the whole body, the various consciousnesses constitute the overall consciousness.

As the *Daode jing* explains, memory is but a component of consciousness. At the same time, different types of form have different capacities for consciousness and cognition. Whereas a human being, for example, is farther from the original Dao but can become cognizant and thus fully conscious of Dao, a mineral, plant, or animal is closer to the original Dao but cannot become cognizant of Dao and experience it fully. Dao, of course, desires to experience itself—for form to become the original consciousness itself. That is why being human is so significant.

Akin to consciousness and cognition, attention and intention are a second important yin-yang dimension for the experience of Dao. Dao is ultimate Attention, while the One is transcendent Intention for that ultimate Attention to be meaningful. The yin-yang Two is then an attention-intention dynamic, and finally the yin-yang Two and embodied Three are a third attention-intention dichotomy. The experience of Dao requires an attentiveness, but for that attentiveness to be meaningful there must be an intentionality.

At the level of yin-yang Two, the attention is a link to the transcendent One and ultimate Dao. At the level of embodied Three, form is a matter of intentionality linked to the yin-yang Two as attentiveness. A. C. Graham provides a formula for the state of total awareness, so that every detail of a situation participates in directing a person's response: "skillful spontaneity is action that arises naturally from a state of total and penetrating attention to the situation at hand" (1989, 29). This total awareness is the perfect attunement of a person with his or her surroundings.

Beyond consciousness and cognition as well as attention and intention, intuition and logic form a third yin-yang dimension for the experience of Dao. The first chapter of the *Guiguzi* 鬼谷子 (Master of the Demon Valley) discusses the relation between wisdom and categories, which are akin to intuition and logic. Dao itself is an immediate Intuition, while the One is a corresponding proximate Logic or Logos. At the level of yin-yang Two, the yin is an immediate intuition and the yang is the step-by-step logic. Finally, at the level of form, the yin-yang Two is an intuition which encompasses a logical component (together an inner knowing), while the embodied Three is a logical three-dimensional reality. As the holistic explanation demonstrates, intuition is primary, although for that intuition to be meaningful, the step-by-step logic must also exist.

As explained earlier with respect to the Middle Way, the Daoist cosmology and metaphysics provide for the yin-yang dynamic to be harmonious and thereby return to the transcendent One and attain the ultimate Dao. In such a state, with the primary yin aspect stable and the secondary yang aspect secure, the two sides support and become one another. This convergence takes place at every layer of form, not just consciousness and cognition but also consciousness and the particular form itself.

It also occurs at every scale of form, including the cellular and muscular scales in human beings. In such a state, the overall form experiences a high degree of integration, both within and without—with the larger universe. Attention and intention merge, and a person intuits the environment accurately and processes information efficiently. In such a state, human beings become capable of skillful spontaneity. In ancient Daoist thought, this state of effortlessly comporting oneself in the world, is marked by fine responsiveness, skill, and enjoyment, is referred to as effortless action.

Whereas the cosmology and metaphysics of effortless action are best expressed in the *Daode jing*, a key example of skillful spontaneity appears in the *Zhuangzi*. Its third chapter tells the story of Cook Ding and the duke:

in response to the latter's praise for his butchering skills, Cook Ding responds that he is "good" because of the Middle Way course (Dao), which is "something that goes beyond mere skill." Cook Ding then relates that to acquire his ability he had looked at oxen for three years and was still unable to see "all there was to see in an ox." This line conveys that the Middle Way requires more than just seeing with one's eyes or understanding an object analytically.

He continues that he now encounters the ox "with spirit rather than scrutinizing it with the eyes." Here, "spirit" could be understood as consciousness, the yin property linking a person to Dao. He quiets the cognitive (*guanzhi zhi* 官知止) and lets "the promptings of spirit begin to flow" (*shenyu xing* 神欲行), relying on "heaven's properties" (*tianli* 天理), which indicate the joints and openings in the ox. He never has to cut through coarse material, "for the joints have spaces within them, and the very edge of the blade has no thickness at all. When what has no thickness enters into an empty space, it is vast and open with more than enough room for the play of the blade" (Ziporyn 2009).

Because Cook Ding goes along with emptiness and does not contest it, his "knife is still as sharp as if it had just come off the whetstone, even after nineteen years." As a holistic explanation provides, at their deepest points the knife and ox are in a state of latency or nonbeing (*wu* 無) as opposed to thinghood (*wu* 物), and thus there is no friction or conflict between the two forms. Even with his skill, he still sometimes comes to a "clustered tangle" in the ox carcass where it is "difficult to do anything about it." In such instances, he restrains himself as if terrified, until his seeing (i. e., cognition) comes to a complete halt.

At this point, "activity slows, and the blade moves ever so slightly. Then all at once, I find the ox already dismembered at my feet like clumps of soil scattered on the ground. I stand there gazing at my work arrayed all around me, dawdling over it with satisfaction." In this particularly challenging experience, the cook goes into an even deeper state of effortless action than usual, where the self along with time and space are altered. As this sentence also communicates, the experience is highly enjoyable.

In his classic, *Zen in the Art of Archery*, Eugen Herrigel provides a contemporary account of this state. He relays the following archery lesson with his master:

> Herrigel: How can the shot be loosed if 'I' do not do it?
> Master: 'It' shoots.
> * * *

Herrigel: And who or what is this 'It'?
Master: Once you have understood that you will have no further need of me. (1953, 51-52)

In ancient Daoist thought, the "it" is consciousness as the origination of all form and power, a universal force running throughout reality. In a state of effortless action, a person is both accessing and experiencing this ultimate reality.

For the Middle Way to be meaningful, there must also be the possibility for a deviation from it. In such a discordant state, the yin aspect is unstable and the yang aspect is insecure. The two aspects then deny and exhaust each other, and reality is experienced as a struggle. In discord, a person perceives and thinks of the individual self as being separate and in conflict with other forms, or bodies. In short, the person becomes self-conscious in an unhealthy way.

Attention and intention are disrupted, so the person has difficulty staying attentive and using intention to accord with reality. In such a self-conscious state, moreover, the mind is considered important, but the mind and body relationship cannot be discerned. Intuition turns into an impulsiveness and logic devolves into a contradictory dialectics (Pratt and Zhao 2019).

Finally, the discordant consciousness and cognition manifests on not just a corporeal but also a muscular and even a cellular level. With consciousness in conflict with cognitive matter, and even with physical matter, every task, not to mention a complex task, requires great thought and strenuous effort and the outcome remains sub-optimal (*Daode jing*, chs. 3, 46, 53). The conflict can even induce mental and physical illness (*Huangdi neijing*, ch. 1). The third chapter of the *Zhuangzi* begins by noting that the Middle Way, described as a waterway course (*ya* 涯), does not prioritize knowledge (*zhi* 知). The next two sentences point out how deviating from the Middle Way is dangerous, and even more so for those who prize knowledge (*zhizhe* 知者) in such a situation. The fate of this path is to follow a central meridian (*du* 督), which allows for the preservation of the body, manifestation of a full life, flourishing of one's companions, and maximization of one's years.

In the story about Cook Ding and the duke, the former remarks that butchers who do not accord with the Middle Way course, i. e., access the state of effortless action, either slice the ox carcass and have to change their blade once a year or hack at the carcass and must change their blade once a month. On the other hand, by accessing effortless action, he has been "using the same blade for nineteen years, cutting up thousands of oxen, and yet it is still as sharp as the day it came off the whetstone."

As the classics show, cognition is never simply neutral, nor is the world simply objective. Reality depends primarily on consciousness and only in a harmonious state is the world perceived, conceived, and acted upon in an accurate and effective manner. In such a state, a person accessing effortless action accords with the Middle Way and may return to the transcendent One and ultimate Dao. In a discordant condition, on the other hand, the person experiences the world as contradictory and frustrating. In any activity, be it butchering or archery, the conscious person may go within and, in doing so, reconnect with effortless action to rise above the ostensible contest at hand. The experience of reality involves not just the individual self (form) but also the other (forms). With the multiplicity of forms, from the smallest to the largest, within oneself and in an environment, there is the possibility of conventional time and space. A significant part of form, as well as either a harmonious or a discordant experience, involves time and space.

Time and Space

In addition to explaining consciousness and cognition as well as form, Daoist cosmology and metaphysics elucidate time and space. A temporal dimension and spatial context are both necessary in the emergence of form, and Daoism shows how time and space have an immediate circular as well as a proximate linear aspect. As already noted, Dao as ultimate reality is beyond all conventional attributes, including time and space, and could be described as the ultimate immediate or circularity; in colloquial terms, Dao is the Here and Now.

In contrast to Dao, the One is the transcendent proximate or linearity. Like the One, this proximate or linearity exists beyond any sense of separate form. Because Dao and the One are both the same and different, at the level of the transcendent One the circular and linear are intertwined. Next, mirroring Dao and the One, the yin-yang Two is an immediate-proximate or circular-linear dynamic, with the yin as the immediate or circular aspect and the yang as the proximate or linear aspect.

At the level of form, the yin-yang Two reflecting Dao is a hidden immediate or circular essence, while the embodied Three reflecting the One is a manifest proximate or linear counterpart. As this holistic explanation shows, even at the level of form, time and space include both the immediate circular here and now leading to the transcendent One, as well as the proximate linear past and future, near and far for that elevated experience to be meaningful. Like the Dao, the circular aspect of time and space can only be pointed to and perhaps is best understood as being in a complementary relationship with the relative aspect of time and space. In accordance with this cosmology and metaphysics, the Daoist classics emphasize staying in the moment and observing the right time to perform a given task, rather than simply the linear passage of time (Raphals 2019).

The immediate and proximate aspects of time and space correspond to absolute and relative dimensions. Again fitting the absolute and relative into the holistic account of reality, Dao is the absolute, while the One is the corresponding relative. At the next level, the yin-yang Two is an absolute-relative dynamic. Then, at the level of form, the yin-yang Two is an absolute aspect (containing both absolute and relative components) and the embodied Three is a relative aspect.

Applying the absolute and relative to time and space at the level of Dao and the One, the ultimate circularity of time and space is absolute, while the corresponding transcendent linearity is relative. At the level of yin-yang Two, the yin circular aspect of time and space is absolute, while the yang linear aspect of time and space is relative. At the level of yin-yang Two and embodied Three, the yin-yang Two as a yin circular aspect is a further absolute aspect, while the embodied Three as a yang linear dimension is a further relative aspect.

Albert Einstein and other physicists were correct in that, contrary to Isaac Newton's original conclusions, linear one-dimensional time and three-dimensional space are relative. Modern physicists, however, have perhaps not yet grasped that an absolute aspect of time and space still exists, and that time and space, as with form, are layered, deriving from an ultimate absolute.

As Daoism shows, moreover, linear time and space are not distinct properties apart from form but instead, arise with form. Einstein and other physicists have understood that linear time and space are interrelated (Einstein named this phenomenon "spacetime") and always relative to form, but they have wrongly thought of spacetime as a separate four-dimensional phenomenon (i. e., three dimensions of space and one dimension of time).

In a holistic account of reality, form, whether on a granular or galactic scale, does not exist against a backdrop of "spacetime," but with time and space. As with form itself, moreover, time and space again are layered, from an ultimate and then transcendent layer of reality to an intermediate energetic layer to a manifest ostensibly material layer. Reality is one integrated and essentially circular process. To the extent that Einstein thought "spacetime" existed as a backdrop to material form and was relative as opposed to absolute, the term "spacetime" is a mischaracterization of reality. As just noted, time and space along with form are layered and always exist together. Rather than invent another term, such as "form-spacetime," the holistic account simply uses the individual words respectively.

Logically speaking, a complete explanation of time and space could not be any other way than as Daoism provides. Any other explanation of form would leave a gap that could not be explained and then also be contradictory. If one assumes a linear past-to-future sense of reality, for example, the beginning and end of this linear time frame cannot be explained. But if one sees that the present (here and now) is all that ultimately exists, and the past and future, near and far, are only for that experience, the beginning and end can be explained. Similarly, if one assumes a "thing/nothing" dualistic description of reality, empty space makes no sense; but if one assumes the two "no thing/thing" as forever interwoven with nothingness as ultimate reality, conventional space does make sense.

The experience of time and space are a function of not just form (i. e., the embodied Three) but, even more importantly, consciousness and then cognition (i. e., the yin-yang Two). First, form is primarily consciousness and then cognition, rather than simply physical matter. Even an experience like memory involves primarily an energetic consciousness and cognition, and only secondarily a material manifestation such as brain or muscular tissue. Second, time and space are a combination of the absolute and relative, akin to consciousness and cognition, and thus the experience of time and space depends on consciousness and then cognition. Finally, since linear time is relative and only an artifice, linear time and space depend on consciousness and cognition. A person has an intuitive or "felt" sense of time and space and then uses logic (or dialectics) to explain time and space. Depending on consciousness, time may seem slow or fast, while space appears large or small. At the same time, time and space can be appreciated as a holistic circular-linear phenomenon or in confusion understood as a reductionist linear reality.

People frequently report the experience of time dilation in everyday activities such as taking a walk or reading a book. Einstein is said to have asked his secretary to explain relativity to inquirers simply through the following comparison: "An hour sitting with a pretty girl on a park bench passes like a minute, but a minute sitting on a hot stove seems like an hour" (Sayen 1985, 130). Like Cook Ding, people also regularly experience extreme time dilation in challenging situations. In a threatening situation like a car crash, researchers usually attribute time dilation to a neurological process that involves the release of hormones and neurological transmitters (e.g., Stetson, Fiesta and Eagleman 2007). According to this explanation, information is transmitted through a physical process at a different rate than usual—a so-called reptilian "fight or flight" response. Time dilation is sometimes also thought to be an "artifact of memory" rather than a trait of real-time experience (Stetson, Fiesta, and Eagleman 2007).

According to the holistic account of reality, however, the experience of time dilation depends primarily on consciousness, though time is processed by cognition and manifested as form. In a threatening situation, a person may become highly conscious and thus time is experienced as more absolute and circular than it normally would be. At such a moment the linear aspect of time may appear to slow down or even stop. The more conscious one becomes, the more one accesses the circular absolute aspect of time, the more linear time may slow down. Again, the experience of linear time is but a relative phenomenon and only for the experience of the immediate circular aspect of time.

Many people have also reported differences in the experience of space, sometimes including an out-of-body experience (Metzinger 2005, 57-84). Like with time dilation, these reports are consistent across historical time periods as well as divergent cultural contexts (Blanke 2004, 1114-15). These experiences may also involve a challenging situation, as in the previous car crash example, but are also again reported in activities like meditation, sleep, and sports, each of which can involve a deep state of consciousness (Metzinger 2005, 59; Alvarado 2000, 183-218). Out-of-body experiences are normally explained as dissociative experiences arising from different psychological and neurological factors, essentially as something made up by the brain (2000, 184).

As with the explanation of time, however, the holistic account of reality shows how the different experiences of space are related to consciousness and the absolute circular aspect of space. In a highly conscious state, people become aware and connected to the absolute aspect of space. In such a state, conventional boundaries between human and other forms may decrease in importance or cease to exist at all. People may also have an out-of-body experience where they become the larger consciousness and look back at their relativistic three-dimensional form. Depending on their cognition, they may understand the absolute aspect of space as true and the relative aspect of space as but an artifice. Otherwise, they may regard the unconventional experience of space as something the brain simply conjured up.

Time and space are a combination of both the circular absolute and the linear relative and are ultimately a matter of consciousness. In a harmonious condition, people can experience the linear aspect of time and space, including the past and future or near and far, as being but for the circular aspect, which is the immediate here and now. Attention and intention complement each other, and they have the intention to stay in an attentive spot. With consciousness stable and cognition secure, they can also intuit events accurately and processes information efficiently.

As consciousness and cognition exist also at the muscular and even at the cellular level, people can act effortlessly and may feel as if they have endless capacity. In such a state, time slows down, though when compared to an unconscious state, it may seem to have passed quickly. Similarly, space grows large but, when compared to a regular state, may appear to be small. In a state of effortless action, for example, a samurai may distinguish the scratch marks on an opponent's slicing sword and, based on "muscle memory," duck or deflect the sword before countering with an effortless maneuver to disarm the attacker (e. g., Ueshiba 2012, 116-29).

In a discordant state the circular and linear aspects of time and space are experienced as contradicting, chasing, and depleting each other. In practice, people may experience this conflict as an uncomfortable tension between a need to be in some other place and a fixation on some point in the past or future, rather than being present to deal with an existing challenge. With attention and intention also in conflict, the mind runs off to solve the perceived problems, or wanders off to escape the feeling of alienation and insecurity.

In some cases, people may even fall into a downward spiral of thinking about their personal problems. In confusion, time and space are considered linear and, as Newton once concluded, this linear time and space are thought of as absolute. People neither intuit events accurately nor process information efficiently. Finally, because in discord the conflict takes place at even the muscular and cellular levels, they are unable to act, either in a temporally or in a spatially accurate manner. All action requires great thought and effort, the opposite of effortless action, and thus becomes exhausting. The unskilled swordsman cannot disarm an opponent, let alone defend an attack against him or herself.

As Daoism's holistic explanation of reality demonstrates, time, space, and form are an integrated reality. Time and space arise with form and like form have both yin and yang aspects. These aspects include the circular and linear as well as the absolute and relative. Ultimately, time and space are the here and now; and the near and far, as well as the past and future, exist only for this immediate experience. Though linear time and space are usually considered as fixed constructs, time and space are often also experienced in subtle and even transcendent ways. Western researchers, especially in the areas of psychology and neuroscience, have studied positive reports of time and space that typically also involve a heightened consciousness, a quiet cognition, attention and intention, and even muscular skill. This research supports, as the Daoist texts elucidate, that the full experience of form as well as time and space is possible even in ordinary pursuits.

The Flow Experience

The central experience of effortless action, with its altered states of time and space, has been noted by modern scientific researchers, especially in connection with positive psychology. In the parlance of psychology, effortless action is referred to as "flow." Coined by Mihaly Csikszentmihalyi, flow has been defined as the mental state of operation in which a person performing an activity is fully immersed in a feeling of energized focus, full involvement, and enjoyment in the process (1975). The modern sense of flow is matter-of-course in Daoism's holistic account of reality—there must be a Middle Way course for returning first to the transcendent One and then to the ultimate Dao. This Middle Way sweet spot is what psychologists are calling "flow."

Csikszentmihalyi recognized that he was not the first person to describe flow, and in his seminal work on the subject even included the *Zhuangzi* story about the duke and the cook as an early characterization of the state (1990, 149-151). He named this sensation "flow" because in his early interviews several people used the metaphor of a current carrying them along to describe their optimal experience. The *Zhuangzi* explains effortless action using the archetype of a waterway course. Dao itself, as a dynamic Middle Way course, also depicts a flow. The concept of effortless action further provides a transcendent sense of flow. Csikszentmihalyi notes researchers have found that when in flow, the individual operates at full capacity, and that the state is one of dynamic equilibrium (2014, 240).

His list of factors accompanying the experience tracks the Daoist sense of effortless action, including shifts in time and space (2014, 240). The experience is "remarkably similar across different leisure and work settings," and "reported in similar terms across lines of culture, class, gender, and age, as well as across cultures" (Nakamura and Csikszentmihalyi 2009, 196). Flow has been noted most often in extreme activities as well as sports and music, but also is found in simple activities like walking and baking. Flow is experienced by not just individuals but also groups, where it may be referred to as "group flow," or simply "being in sync" (Pels et al. 2018).

Flow, with time and space dilation, has often been associated with extreme activities, where serious injury or death are ever-present possibilities. They provide an incentive for going into a heightened state of concentration and efficiency. One of Csikszentmihalyi's books cites rock climbing as an example of a class of deep play flow activities (1975, 74-101). Widely regarded as one of the greatest Formula One drivers of all time, Ayrton Senna described a flow experience of losing his individual sense of self and his race car exceeding normal temporal and spatial limits:

> I was already on pole, . . . and I just kept going. Suddenly I was nearly two seconds faster than anybody else, including my team mate with the same car. And suddenly I realized that I was no longer driving the car consciously. I was driving it by a kind of instinct, only I was in a different dimension. (Orosz 2010)

Effortless action is perhaps most often associated with athletic activities like archery and martial arts, as depicted by Herrigel (1953) and Ueshiba (2012). Jackson and Csikszentmihalyi have studied the experience of flow in conventional athletics (1999), and sports psychologists have noted that the integration of the conscious and subconscious improves body coordination and movement patterns (Palmer 2006, 45-63). Many athletes have recounted the effortless nature of a performance while exceeding expected performance levels. Athletes in the flow zone also report a subtle freedom from self-consciousness and cognition. Former pro baseball catcher Tim McCarver, who reports experiencing "the zone" when catching for record-breaking pitchers like Bob Gibson, describes this unique lack of cognition:

> Many ballplayers think too much. Players like [Bob] Gibson and [Orel] Hershiser seem to have a sort of paradoxical intelligence—one that allows them not to do anything to hinder themselves. It's a sort of intelligence you use almost paranormally. It allows people to do phenomenal things. People who really use their mind—they free it from impeding their activity. (Shainberg 1989)

In the story of Cook Ding, "the knife would whiz through with its resonant *thwing*, each stroke ringing out the perfect note, attuned to the 'Dance of the Mulberry Grove' or the 'Jingshou Chorus' of the ancient sage kings." As this passage notes, effortless action is inherently harmonious, and thus it is unsurprising that flow is commonly reported in connection with musical performances.

Similar to what the cook experienced butchering an ox, musicians in flow report being so totally absorbed in playing their instrument that they lose track of time and are surprised by playing better than they thought they could play (Bloom and Shutnick-Henley 2005, 24). Flow is also connected to the intrinsic enjoyment of playing music. Both improvisational soloists and ensembles may attain a state of aesthetic rapture while playing or singing (Sawyer 2015, 33). Group flow occurs among collaborators in musical performances, because on the level of not just consciousness but also form, performers resonate with one another. One choral director depicts a typical group flow experience as achieving "an especially high level of artistry, accuracy, and focus" (Walters 2016).

The Cook Ding story involves work, and Csikszentmihalyi's book on flow features a chapter entitled "Work as Flow." Some researchers have likened entrepreneurship to extreme experiences and concluded that, with its high risks and high rewards, entrepreneurship should be "approached as a vehicle for optimal human experiencing" (Schindehutte et al. 2006, 349). As the *Zhuangzi* story and modern research indicate, however, a level of effortless action or flow is important for job satisfaction in general (Maeran and Cangiano 2013, 13-26).

In the context of work, people are more likely to encounter the ideal combination of high perceived skill and challenge than they would in leisure activities (Csikszentmihalyi and LeFevre 1989, 817). It may well play a central role in cultivating effortless action. In addition, the harmonious state may also be related to producing positive-sum results for all potential business stakeholders, including employees, suppliers, community members and customers (Pratt 2016).

In the *Zhuangzi*, after Cook Ding finishes the explanation of his skill, the duke exclaims, "Wonderful! From hearing my cook's words, I have learned how to nourish life!" The story shows that there are different levels of effortless action, and although a challenge (a particularly knotty part of the ox) may be necessary to attain the deepest state, people can still experience effortless action in other circumstances (the parts of the ox that are easy to cut). The holistic explanation shows why effortless action in essence and in due accordance with the Middle Way is essential to health and happiness, including a holistic consciousness as well as a full sense of time and space. From Csikszentmihalyi's standpoint as well, flow is a basic component of positive psychology, which of course seeks to maximize human health (1975, x).

The *Zhuangzi* provides a simple explanation of how Cook Ding first cultivated his state of consciousness: by practiced awareness with the ox. Throughout China's long history, meditation and martial arts, including versions of taiji quan, have been practiced to cultivate effortless action and thereby attain a high level of consciousness as well as a full sense of time and space. In one of his books, Csikszentmihalyi has a section entitled

"The Ultimate Control: Yoga and the Martial Arts." He concludes, "In many respects, what the West has accomplished in terms of harnessing material energy is matched by what India and the Far East have achieved in terms of direct control of consciousness" (1990, 103). These cultivation practices involve not just consciousness and cognition but also form. As he also notes, "It makes sense to think of yoga as a very thoroughly planned flow activity" (1990, 105).

Sports coaches and musical directors have recently studied flow in an effort to set conditions to facilitate the state (e.g., Palmer 2006; Walters 2016), while business researchers have sought to apply it in work settings (e.g., Eisenberg et al. 2005). Some modern athletes and business people have even practiced martial arts or meditation to gain access to a state of effortless action in their particular activities. In his autobiography, *A Zen Way of Baseball*, co-written with David Falkner, the great Japanese hitter Sadaharu Oh explains how studying with the legendary aikido master Ueshiba Morihei taught him the critical ability to "wait," which was "the most active state of all" and actually gave him more time to hit the baseball (Oh and Falkner 1985, 168-69). Ray Dalio, founder, chairman and CEO of an over $100 billion investment management firm, credits his business success to a decades-long transcendental meditation practice, as "it helps slow things down so that I can act calmly, even in the face of chaos, just like a ninja in a street fight" (Clifford 2018).

To explain effortless action or flow, Western psychologists and neuroscientists have wrongly tried to use a reductionist model of body and mind as well as an essentially linear sense of time and space, thus failing to do so. Similarly, some scholars have failed to appreciate the transcendent nature of effortless action (Barrett 2011, 685-86). With Daoism's holistic explanation, on the other hand, the sublime experience becomes clear: people in a state of effortless action access an overarching level of consciousness, of which they are a part and which permeates even their muscles and cells. In this state, action is effortless, and both time and space become holistic. Modern psychologists are right, however, that flow is essential for positive physical and mental health. As Daoism indicates, effortless action with the yin and the yang aspects in a complementary condition is the only way to achieve optimal health. Its cultivation can be done through practices of meditation, taiji quan, and yoga. While many flow-inducing practices have traditionally been associated with the martial arts, they are yet also central to the success of athletic competitors and business people.

Conclusion

Daoism's holistic account of reality integrates consciousness, cognition, and form as well as time and space. Time and space have both an absolute circular aspect—the here and now—as well as a linear relative aspect—the one dimension of time and three dimensions of space which allow the absolute here and now to be meaningful. The experience of time and space depends on consciousness, is processed by cognition, and happens through form, at each layer of reality. In a harmonious state of effortless action, form may return to the transcendent One, and time and space are experienced as a deliberate spaciousness. On the other hand, in a disharmonious condition, existence is experienced as a struggle, and time and space are fraught with practical and theoretical contradictions, including how conventional linear time and space arose in the first place.

Further, as the phenomenon of relative or linear time and space depends on consciousness, akin to the absolute or circular aspect of time and space, relative or linear time and space cannot limit the experience of any given form. As this holistic explanation provides, and modern research in psychology substantiates, the transcendent sense of the self in a state of effortless action, including a full sense of time and space, is available to each individual as well as to each group, and can be cultivated and applied in all of life's activities. This positive state, moreover, is essential to optimal physical, mental, and spiritual health for each individual as well as the community as a whole.

Bibliography

Alvarado, Carlos S. 2000. "Out-of-Body Experiences." In *Varieties of Anomalous Experience: Examining the Scientific Evidence,* edited by Etzel Cardena, Steven Jay Lynn, and Stanley Krippner, 183-218. Washington, DC: American Psychological Association.

Barrett, Nathaniel F. 2011. "Wuwei and Flow: Comparative Reflections on Spirituality, Transcendence, and Skill in the 'Zhuangzi'." *Philosophy East & West* 61.4:679-706.

Blanke, Olaf. 2004. "Out of Body Experiences and Their Neural Basis: They Are Linked to Multisensory and Cognitive Processing in the Brain." *British Medical Journal* 329.7480:1414-15.

Bloom, Arvid J. and Paula Shutnick-Henley. 2005. "Facilitating Flow Experiences Among Musicians." *American Music Teacher* 54.5:24-28.

Clifford, Catherine. March 16, 2018. "Hedge fund billionaire Ray Dalio: Meditation is 'the single most important reason' for my success." www.cnbc.com 2018 /03/ 16/ bridgewater-associates-ray-dalio-meditation-is-key-to-my-success.

Csikszentmihalyi, Mihaly. 1975. *Beyond Boredom and Anxiety*. San Francisco: Jossey-Bass.

_____. 1990. *Flow: The Psychology of Optimal Experience*. New York: Harper & Row.

_____. 2014. *Flow and the Foundations of Positive Psychology*. New York: Springer.

_____, and Judith LeFevre. 1989. "Optimal Experience in Work and Leisure." *Journal of Personality and Social Psychology* 56.5:815-22.

Eisenberg, Robert, Jason R. Jones, Florence Stinglhamber, Linda Shanock, and Amanda T. Randall. 2005. "Flow Experiences at Work: For High Achiever's Alone?" *Journal of Organizational Behavior* 26.7:755-75.

Graham, A. C. 1989. *Disputers of the Tao: Philosophical Argument in Ancient China*. La-Salle, Ill.: Open Court.

Herrigel, Eugene. 1953. *Zen in the Art of Archery*. Translated by R. F. C. Hull. New York: Pantheon Books.

Jackson, Susan A., and Mihaly Csikszentmihalyi. 1999. *Flow in Sports: The Keys to Optimal Experiences and Performances*. Champaign, Ill.: Human Kinetics.

Maeran, Roberta and Francesco Cangiano. 2013. "Flow Experience and Job Characteristics: Analyzing the Role of Flow in Job Satisfaction." *TPM* 20.1:13-26.

Metzinger, Thomas. 2005. "Out-of-Body Experiences as the Origin of the Concept of a 'Soul.'" *Mind & Matter* 3.1:57-84.

Nakamura, Jeanne, and Mihaly Csikszentmihalyi. 2009. "Flow Theory and Research" in *The Oxford Handbook of Positive Psychology*, edited by Shane J. Lopez and C. R. Snyder, 195-206. New York: Oxford University Press.

Oh, Sadaharu, and David Falkner. 1985. *Sadaharu Oh: A Zen Way of Baseball*. Visalia, Cal: Vintage.

Orosz, Peter. 2010. "Ayrton Senna on the Very Edge."www.jalopnik.com/ayrton-senna-on-the-very-edge-5452870

Palmer, Roy. 2006. *Zone Mind, Zone Body: How to Break Through to New Levels of Fitness and Performance – By Doing Less!* Lagos: Academy Press.

Pels, Fabian, Jens Kleinert, and Florian Mennigen. December 31, 2018. "Group flow: A scoping review of definitions, theoretical approaches, measures and findings." *PLoS ONE* 13.12. www.doi.org/10.1371/journal.pone0020117

Pratt, Joseph. 2016. "So Said Laozi: How Daoism Explains 'Creating Shared Value'." www.ssrn.com/abstract=284338.

_______, and Chenting Liu. 2018. "道家还原论宇宙观中的谬论: 以老庄思想和《黄帝内经》为中心的分析." www.ssrn.com.abstract=3229541.

_______, and Yingnan Zhao. 2019. "A Daoist Critique of Dialectics and Why It Matters." www.ssrn.com/abstract=3140946.

Raphals, Lisa. 2019. "Time, Chance, and Fate in Early Daoist Texts." Keynote Address at the 13th International Conference on Daoist Studies.

Sawyer, Keith. 2015. "Group Flow and Group Genius." *The NAMTA Journal* 40.3:29-52.

Sayen, Jamie. 1985. *Einstein in America: The Scientist's Conscience in the Age of Hitler and Hiroshima.* New York: Crown Publishers.

Schindehutte, Minet, Michael Morris, and Jeffrey Allen. 2006. "Beyond Achievement: Entrepreneurship as Extreme Experience." *Small Business Economics* 27.4-5:349-68.

Shainberg, Lawrence. April 9, 1989. "Finding the Zone." *The New York Times Magazine*. www.nytimes.com/1989/04/09/magazin/finding -the-zone.html.

Schmidt, Richard A., and Timothy D. Lee. 2005. *Motor Control and Learning: A Behavioral Emphasis*. Windsor, Ont.: Human Kinetics.

Stetson, Chess, Matthew P. Fiesta, and David M. Eagleman. 2007. "Does Time Really Slow Down During a Frightening Event?" *PLoS* ONE. www.doi.org/10. 1371/ journal. pone.0001295.

Walters, Christopher M. 2016. "Choral Singers 'In the Zone': Toward Flow through Score Study and Analysis." *The Choral Journal* 57.5:8-19.

Wang, Robin R. 2012. *Yinyang: The Way of Heaven and Earth in Chinese Thought and Culture*. New York: Cambridge University Press.

Ueshiba, Morihei. 2012. *The Secret Teachings of Aikido.* Translated by John Stevens. Tokyo: Kodansha International.

Ziporyn, Brook. 2009. *Zhuangzi: The Essential Writings with Selections from Traditional Commentaries*. Indianapolis: Hackett Publishing.

Temporality in Laozi and Plotinus

ANDREJ FECH

The writings attributed to Laozi and Plotinus were created in profoundly different cultural and philosophical environments and several centuries apart. Yet, the philosophical similarities between them are intriguing and have rarely been studied. They include the centrality of the notion of the "One," the ineffability of its nature, the gestation of the physical world from it, the outline of several distinct stages in this process, and, finally, the prominence of the notion of return. It is my contention that comparing these two thinkers is also conducive to understanding their respective positions with regard to the phenomenon of time.

When discussing the *Laozi*, I consider early versions of the text, discovered in recent decades. The earliest was excavated in Guodian 郭店 and presumably circulated in the southern state of Chu 楚 toward the end of the fourth century BCE (Allan and Williams 2000, 120). Furthermore, there are three Western Han (206 BCE-6 CE) versions, two (A and B) unearthed at Mawangdui 馬王堆 and one belonging to the Peking University collection. They shed light on the contents and the structure of the work during early imperial times, before it was cast into its transmitted form (Liu 2017). Plotinus' writings are quoted based on the edition used by A. H. Armstrong in his full translation of the philosopher's works published between 1966 and 1988.

Temporality in the *Laozi*

Temporal characteristics play a crucial role in the *Laozi*. They serve as the ultimate criterion to determine whether or not a certain phenomenon complies with central values and notions of the text, of which the Way or Dao is most emblematic. Thus, the all-too-brief duration of any occurrence in the world—such as human life, natural events, specific actions, and more—is seen as resulting from disobeying these values (chs. 9, 23, 24, 30, and 55). *Laozi* 55 expresses this issue with utmost directness: "What is not the Way comes to an early end" (*budao zaoyi* 不道早已).

On the other hand, phenomena that comply with the norms promulgated by the text are said to last long (chs. 7, 16, 33, 44, and 59). Even the legendary author of the *Laozi*, the eponymous archive keeper of the Zhou capital, is reported in several accounts, beginning with the *Shiji* 史記 (Records of the Historian), to have lived for several centuries because of "his sanctity and of his harmony with the Dao" (Seidel and von Falkenhausen 2008, 142).[1]

Against this background, it is not surprising to find the Way itself associated with permanence and endurance throughout the text. In some cases, this idea is expressed directly, such as in chapter 16, stating: "To be the Way is to endure" (*dao nai jiu* 道乃久). In others, it is implied variously, so that the Way is called the "forefather of God" (*di zhi xian* 帝之先) (ch. 4) or described as having been born "before heaven and earth" (*xian tian di sheng* 先天地生) (ch. 25). In addition, chapter 14 addresses the Way's antiquity, while at the same time stressing its availability in the present moment:

> Hold fast to the way of antiquity
> In order to keep in control the realm of today.
> The ability to know the beginning of antiquity
> Is called the thread of the Way.[2]

Moreover, the Way is frequently discussed by means of weaving metaphors, such as the "thread of the Way" (*daoji* 道紀) in this chapter. This metaphor stresses the linear continuity of the object described (Loewe 1995, 312). Similarly, the use of reduplicative binomes with the semantic element "silk" (*mi* 糸), such as *mianmian* 綿綿 (ch. 6) and *shengsheng* 繩繩 (ch. 14), also underscores the uninterrupted continuity of its existence (Chen 2006, 99, 127). Therefore, as most academic papers dealing with Daoist concepts of time agree, the Way is eternal in its nature (Dy 1996, 255; Jhou 2020, 585). This is consonant with the observation that the Way, while be-

---

[1] A. C. Graham assumes that Laozi's alleged longevity was only a byproduct of the political move carried out by his admirers to earn patronage from the rulers of the superpower of Qin. They identified Laozi, otherwise known as the elder contemporary of Confucius (551-479 BCE), with the Grand Historiographer Dan 儋, who prophesied the ascendancy of Qin over Zhou around 374 BCE, thus extending his lifespan to several centuries (1998, 32-33). Regardless of whether or not this conjecture is correct, already in the early Han, Laozi was seen as a "practitioner of longevity" (Kohn 1998, 42). Studies on some prominent disciples of Laozi support this view (Fech 2015, 241).

[2] 執古之道, 以御今之有. 能知古始, 是謂道紀 (Lau 2001, 20-21.)

ing the origin of the world, "remains as a permanent background condition of existence" (Lewis 2006, 24).

At the same time, the reasons for the perpetual existence of the Way are rarely studied. Indeed, the matter seems confusing. At first glance, the eternal characteristics of the Way seem to be a direct consequence of it shapeless, nameless, changeless, and motionless nature, stated among others in *Laozi* 25: "It stands alone and does not change" (*du li er bugai* 獨立而不改).[3] However, such understanding is immediately challenged in the following lines, where the Way is said to "go past" (*shi* 逝), "go far" (*yuan* 遠) and "return" (*fan* 反). In addition, "return" is defined as the very motion of the Way (*dao zhi dong* 道之動) in *Laozi* 40. To resolve this contradiction, I first discuss the two notions that are closely related to the temporal qualities of the Way, namely, *chang* 常 and *heng* 恒.

Chang and *Heng*

The term *chang* figures prominently in the received versions of the text. It appears in eighteen different chapters and is often interpreted as "eternal" or "eternity." Accordingly, the famous opening lines of the transmitted *Laozi* say: "The Way that can be *dao*-ed is not the eternal Way."[4] However, this interpretation is problematic for a number of reasons.

To start with, as the early versions unanimously demonstrate, in most cases where the received text reads *chang*, the original reading was *heng*. The complete replacement of *heng* with *chang* was most likely due to the practice of observing naming taboos, in this case, the personal name Heng of Emperor Wen (203-157 BCE). From the earliest appearances of the character *heng* in written sources, it seems that it originally indicated a moon crescent (Allan 2003, 269-70; Wang 2019, 23). Given the incessant alternation of the moon's waxing and waning, the notion seems to have signified a constant pattern of change rather than the continued existence of any given object. Therefore, translating it as "constant," as commonly done, appears appropriate.

---

[3] In the transmitted version, this is followed by the statement about the incessant cyclical movement of the Way: *zhouxing er budai* 周行而不殆 (Lau 2001, 36-37). However, none of the three earliest versions (Guodian, Mawangdui A and B) contains this sentence.

[4] 道可道非常道. For a list of translations and a related discussion, see Wohlfart 2001, 45-50. It is noteworthy that this interpretation of *chang* was promoted as early as the "Jie Lao" 解老 chapter of the *Hanfeizi* 韓非子. For more, see Queen 2013, 241.

In the early versions of the *Laozi*, *heng* mostly appears either as an adjective or, more often, as an adverb (Lau 2001, 162; Liu 2006, 514-15).[5] As an adjective, *heng* serves as one of the few predicates of the central notions of the Way and "virtue" (*de* 德). In other early cosmological accounts, the Way—when predicated by *heng*—stands for a recurring pattern in the incessant flux of change.[6] In the *Laozi*, most adjective usages of *heng* appear in the context of "return," such as: "When your constant virtue is complete, you will return to the state of uncarved wood" (ch. 28).[7] Therefore, the characteristic resistance of the "constant Way" to becoming the subject of *dao*-ing as expressed in the transmitted *Laozi* 1, may well have to do with *heng*'s association with "return." This is to say, the "constant Way" differs from the "common" ways that are subject to *dao*-ing through its constant return to the nameless and formless.

Unlike *heng*, *chang* features in the early copies only as a noun, appearing in three passages that correspond to the received chapters 16, 52, and 55 (Chen 2017, 334-35). *Chang* does not serve as a predicate to any other notion and appears to be a philosophical principle in its own right. Indeed, the text claims that even possession of the Way is predicated on whether or not one knows *chang* (ch. 16). While the *Laozi* does not provide an exact definition of *chang*, a look at other early texts reveals that this notion mostly referred to unvarying, regulatory principles in the operations of nature, state or human body.[8] In view of this, it appears fitting to translate *chang* as "norm." Characteristically for the *Laozi*, the "normative" principle re-

---

[5] A possible exception to this is the definition of *heng* given in the Mawangdui equivalent of the transmitted chapter 2: "The mutual filling of high and low, the mutual harmony of tone and voice, the mutual following of front and back— these are all constants." 長短之相刑也, 高下之相盈也, 意聲之相和也, 先後之相隋, 恆也. (Henricks 2000, 190-91).

[6] For instance, the "Lunyue" 論約 (Discourse on the Quintessential) from the excavated manuscript corpus *Huangdi sijing* 黃帝四經 reads: "Then he will observe the constant Way of heaven and earth (*tiandi zhi heng dao* 天地之恒道)." Cf. Chang and Yu 1998, 140. In this particular case, the constant Way refers to the alternating pattern of *wen* 文 and *wu* 武 exhibited by heaven and earth.

[7] 恒德乃足, 復歸於樸. (Lau 2001, 42-43).

[8] The *Huangdi sijing* text "Daofa" 道法 (The Way and Law) defines the "constant *chang*" (*hengchang* 恒常) of "heaven and earth" as the "four seasons" (*sishi* 四時),"day and night" (*huiming* 晦明) etc. For more, see Chang and Yu 1998, 104. For the meaning of *chang* in politics, see the stele inscription of Qin Shi Huang 秦始皇 as recorded in the *Shiji* (Nienhauser 1994, 152). For *chang* in bodily functions, see the *Huangdi neijing* (Unschuld 2016, 125).

ferred to as *chang* appears to be closely connected to return (Chen 2016, 3-6; Chen 2017, 342-43).

Thus, rather than signifying temporal endurance, *heng* and *chang* emphasize the normativity of return. Before analyzing what return actually entails for the Way and how it is connected to the latter's permanence, I turn to the *Laozi*'s cosmogonic accounts. There, I focus on the notion of "harmony" (*he* 和). Because, together with return, it is the only notion in the *Laozi* to be defined as *chang*. [9]

Harmony

There are several metaphors describing the emergence of the physical world in the *Laozi*. While the metaphor of birth is preeminent, accounts of movement, growth and differentiation also play an important role, emphasizing different aspects in this complex generative process (Fech 2018, 2). In most cases, the Way is described as a caring entity providing a constant nourishment and protection for the ten thousand things in a most unassuming way (ch. 51). Due to this sympathetic attention, the development of things follows a distinct pattern: Beginning as small, weak and soft entities, they grow to gradually embody such characteristics as big, strong and hard (chs. 63, 64). For the sake of convenience, I follow Angus Graham in designating the first group of (small) qualities "B" and the second group of (big) qualities "A" (1989, 223).

In some cases, the *Laozi* provides a more complex picture of world origination. In chapter 40, for instance, the ten thousand things are held to emerge from "being" (*you* 有), which, in turn, is generated by "non-being" (*wu* 無).[10] And, according to *Laozi* 42, the world emerges from the Way via several distinct stages, referred to by the numerals "one," "two" and "three":

> The Way generates the One.
> The One generates the Two.
> The Two generates the Three.

---

[9] All early copies of *Laozi* 55 define "harmony" as *chang* (*he yue chang* 和曰常). The transmitted text reads instead "to know the harmony is called *chang*" (*zhi he yue chang* 知和曰常). Based on the more frequent appearance of *he* as an independent notion in the early copies, Liu Xiaogan concludes that it played there a more prominent role than in the received text (2017, 107-09).

[10] In the Guodian manuscript, the corresponding passage reads: "The things in the world come from being, [and] come from non-being" (*tianxia zhi wu sheng yu you sheng yu wang* 天下之物生於又生於亡). For more, see Bai 2008, 342-43.

The Three generates the ten thousand things.
The ten thousand things carry yin on their backs and embrace yang.
By blending the vital force [of yin and yang] they obtain harmony.
What is hated by the people is to be solitary, desolate, and hapless,
Yet kings and dukes refer to themselves in this way.[11]

The exact meaning of the three numbers has been subject to radically different interpretations over the millennia.[12] Likewise, the significance of this passage and its relation to other cosmogonic parts of the *Laozi* has been judged very differently.[13] Nevertheless, the passage is informative for several reasons.

First, there are good reasons to assume that time comes into play at the precosmic level, prior to the appearance of the ten thousand things.[14] This would be in line with several other cosmogonic accounts, both transmitted and recently discovered. There, temporality, either as the "four seasons" (*sishi* 四時) or as part of the compound "space-time" (*yuzhou* 宇宙)[15], appears prior to the completion of the physical world. The almost uniform emphasis of time's antecedence to the world goes against the popular opinion that there was no concept of abstract time in ancient China.[16] So, while the main notion for time (*shi* 時) originally meant "season" or "time-

[11] 道生一, 一生二, 二生三, 三生萬物. 萬物負陰而抱陽, 沖氣以為和. 人之所惡, 唯孤、寡、不穀, 而王公以為稱. (Lau 2001, 62).

[12] Wang Bi discusses this passage in terms of his idea of the Way as "nothingness" (Wagner 2003, 266-67). Heshang gong interprets "two" as yin and yang and "three" as "heaven, earth, and man" (Erkes 1945, 196). Cf. Chan 1991, 48, 125.

[13] Girardot claims that "the mode of creative function established by the harmonious interaction of duality" during the "precosmological" period depicted in this passage constitutes the very reason for the Way's "creative activity and enduring presence in the world" (1989, 55). On the contrary, Perkins assumes that this line "sits uneasily" with the other, allegedly more fundamental, cosmogonic account of the text, in which "things emerge from *dao* or no-being (*wu*)" (2015, 217).

[14] Kim, for instance, holds that "the number "three" nests in the Chinese perception of time" (2012, 40).

[15] For the concept of the "four seasons" in the Guodian manuscript *Taiyi shengshui* 太一生水 (The Great One Gives Birth to Water) and the *Huangdi sijing* text "Guan" 觀 (Investigation), see Wang 2015, 7. For "time and space" in the transmitted texts, see *Huainanzi* 淮南子 (He 1998, 166).

[16] Accordingly, time concept in China "is always considered a property of life activity, creativity, generation, and transformation of individual things" (Cheng 1974, 156). Moreover, time and space, alongside other aspects of the "emergent order," "cannot stand as universal principles, as necessary, a priori conditions that give us a single-ordered world" (Hall and Ames 1995, 186).

liness" and entailed a specific set of actions, as in "doing something at the appropriate time" (Allan 1998, 11-12), the *Laozi* and other early texts show that ancient Chinese also conceived time as transcending the particularities of concrete life situations (Harbsmeier 1995, 49).

Furthermore, the above passage is significant for its treatment of "harmony" (*he* 和) as a balanced state of the two generative forces of yin and yang. The final sentence about what the people hate presents the rulers' practice of referring to themselves with demeaning predicates. It is sometimes believed to have originally belonged to another chapter (Chen 2006, 238). However, to me, it exemplifies the working of harmony on the political level. That is, a ruler, who is the embodiment of greatness and multiplicity in a state since he rules over the "ten thousand" people, obtains a balanced position by publicly associating himself with the smallness of the number "one," by humbly presenting himself as solitary and lonesome. A similar idea appears in chapters 39 and 66, which deal with the necessity for a (sage) ruler to display humility. The latter views this practice as a key for winning the unwavering support of the people.

In Graham's parlance, for a head of state, "harmony" is obtainable in the movement from qualities A (greatness) to qualities B (smallness), which reverses the usual development pattern of things. Significantly, the qualities in question belong to the different realms of social status (A) and public appearance (B). This implies that none of them needs to be abandoned or even diminished to obtain harmony. On the contrary, by resigning his position, a humble ruler would destroy the harmonious relationship between his status and outer appearance. The investigation in the previous section showed that harmony and return were the only terms in the *Laozi* defined as "norm." In the following section, I demonstrate that there are, in fact, great similarities between the harmonizing strategies of rulers and the return of the Way.

Return

The *Laozi* uses a number of terms to express the idea of return, most importantly *fan* 反, *fu* 復, and *gui* 歸.[17] The subject of the returning motion can

---

[17] The notion of return features prominently in a number of early works in a variety of meanings, as distinct from the *Laozi*. In the cosmogonies of the *Hengxian* 恒先 (Constant Primacy) and *Taiyi shengshui*, it relates to the "generative process" (*sheng* 生). Some scholars distinguish between three ways of how return and this *sheng* are related. For Xing Wen (2015, 213) return is an "indispensable part of *sheng*," "a new phase of *sheng*," or "a beginning of a new *sheng* process." Others see in *fu*, at

be either the Way (chs. 14, 25, 40), the exemplary human being (chs. 28, 52, 64) or the ten thousand things (chs. 16, 34). In the two latter cases, the destination is usually either the Way or states associated with it. While for the ten thousand things, return implies physical decay and death (Erkes 1945, 154), in the case of an exemplary person it mostly connotes cognition and emulation of the Way.[18] As such, return has utterly positive effects, enabling one to complete the natural duration of a human life without endangering it through strife and excess.[19] The positive outcome of the return is so great that it is even believed to allow a person to escape the cycles of growth and decline—the fate of all living creatures (Kaltenmark 1968, 46; Lau, 2001, xxiii-xxv).[20]

However, when it comes to the Way, it is not immediately clear how it would engage in return, given its portrayal as the starting and ending point of all movement. Not to mention that return here would imply that, at a certain point, the Way becomes alienated from itself. *Laozi* 34 illuminates this:

> The Way is broad, reaching left as well as right.
> The myriad creatures depend on it for life yet it claims no authority.
> It accomplishes its task yet lays claim to no merit.
> It clothes and feeds the myriad creatures yet lays no claim to being their master.
> Forever free of desire, it can be called small;
> Yet as it lays no claim to being master when the myriad creatures turn to it, it can be called great.

---

least, as it occurs in the *Hengxian*, the connotation of "reproduction" and "imitation" (Klein 2013, 214).

[18] Girardot, for instance, views return as embodied in the "stupidity" of the Daoist sage, requiring a "mystic journey back to an individual's fetal origins" (1983, 71).

[19] In chapter 6, the emulation and possession of the Way is told to have the effect of being "free from danger until the end of one's life" (*moshen budai* 沒身不殆). Some scholars determine the function of return as resolving abnormalities and restoring the natural state of things (Wang 2019, 7).

[20] Some passages of the *Laozi* seem to hint at the possibility of continuing existence after physical death. For instance, chapter 33 speaks of longevity as the ability "to die but not perish" (*si er bu wang* 死而不亡). The *Xiang'er* 想爾 commentary interprets this sentence along the lines of rebirth and new existence (Bokenkamp 1999, 135). However, in general, the *Laozi* remains vague on the issue of physical immortality.

> It is because it never attempts itself to become great that it succeeds in becoming great.[21] (Lau 2001, 50)

The Way here reaches both left and right, that is, it embodies opposing principles. Subsequently, the text calls it both "great" (*da* 大) and "small" (*xiao* 小) depending on its relation to the world. It is great because the ten thousand things constantly return to the Way, which the *Laozi* interprets as their enthusiastic submission. The Way effectively becomes their ruler and is thus an embodiment of greatness. However, it does not make any display of great authority, demonstrating instead utter humility by constantly dwelling in the lowest position (chs. 8, 32). This low profile in fact constitutes the Way's return from qualities B to A, creating a harmonic blend between its elevated status and low appearance and earning its designation as "great."

Evidently, this metaphoric return does not imply any physical movement of the Way. It is noteworthy that this idea could not have been formulated without taking into account the world and the cyclical pattern of its existence.[22] In this manner, the Way's enduring existence is not solely predicated upon its primary empty and motionless characteristics but depends on its ability to follow the norm of harmonizing different aspects of its relationship to the world.

Why did the author of the text place such great emphasis on the Way's return? One possible answer is that return is absolutely indispensable for the Way to counterbalance its elevated position as the ruling principle of the world. Then again, by identifying return as the Way's *modus operandi* in the world, the text renders it possible for beings whose existence undergoes cyclical motion to emulate the Way.

In this manner, any objects capable of some form of return can endure in principle. *Laozi* 7 insists that the reason why heaven and earth exist eternally is because they "do not live for themselves" (*bu zisheng* 不自生). Their ability to go beyond themselves can certainly be viewed as a kind of return. In fact, heaven and earth are among the four "great" entities named in chapter 25, for which return is absolutely essential, the other two being the Way and king. The harmonizing strategies of rulers facilitating

---

[21] 大道汎兮, 其可左右. 萬物恃之而生而不辭, 功成不名有. 衣養萬物而不為主, 常無欲, 可名於小；萬物歸焉, 而不為主, 可名為大. 以其終不自為大, 故能成其. The early versions of the chapter show a number of differences to the received text, however, the definitions of the Way's "greatness" and "smallness" remain the same.

[22] Some authors, however, suggest that this returning pattern is established already during the precosmic stage, before the appearance of the "ten thousand things" (Girardot 1989, 55).

their power consolidation and, consequently, the indefinite extension of their reign (ch. 54), can, in fact, be regarded as practical implementation of this imperative.

The World of Plotinus

Considered one of the most prominent representatives of Neoplatonism, Plotinus (205-270) was greatly influenced by Plato's philosophy. In fact, his endeavor is best understood as an attempt to defend Plato's teaching as he understood it against critical voices, specifically those of Aristotle and the Stoics (Gerson 1996, 5-8). This certainly holds true for his main work on issues of time, "On Eternity and Time" (*Ennead* III.9),[23] which provides both the author's deliberations on Plato's account of temporal problems and his critique of other theories. By juxtaposing eternity and time as well as attempting to understand the latter through the former, Plotinus follows Plato's famous precept that time is the moving "image of eternity" (*Timaeus* 37d6-7) while also basing his considerations about temporal issues on his general philosophy. Certain features stand out.

To begin, Plotinus claims that the physical world emerges as the visible image of a more fundamental intelligible reality. In his view, this intelligible realm comprises three basic elements or *hypostasis*: the One (*to hen*), Intellect (*nous*), and Soul (*psuchē*).

The One is the constitutive element of the intelligible world. The source of everything, it is yet ineffable due to its utmost simplicity and unity. According to *Ennead* VI.9, "On the Good or the One," it is "not in place, not in time, but "itself by itself of single form," or rather formless, "being before all form, before movement and before rest; for these pertain to being and are what make it many" (Armstrong 1988, 315).

Intellect is the seat of Platonic ideas or forms. Because thinking and being are identical in Intellect, its being is constituted by its thinking of the ideas (Beierwaltes 1995, 25-27). In contemplating them, Intellect "intuitively" grasps them at once in their entirety, since all ideas exist within it. However, despite the identity of thinker and thought, "the multiplicity of the Ideas means that Intellect does not possess the total simplicity which belongs to the One" (Kenny 2004, 313). Therefore, "plurality in unity" is one of the main features of Intellect.

---

[23] In the arrangement of his disciple Porphyry, who divided Plotinus' works into six groups of nine treatises each (Enneads) (Kenny 2004, 112-13), "On Eternity and Time" is the last, ninth, treatise of the third group (III.9).

Soul is characterized by a greater degree of internal differentiation. Plotinus distinguishes between several parts within it. While its highest segment has the capacity of intellectual contemplation of what is prior to it in the intelligible realm, its lower part, called Nature, has to rely on sensory experience and visual images to somehow grasp the ideas. Physical reality is created for this particular reason. "The duller children, too, are evidence of this, who are incapable of learning and contemplative studies and turn to crafts and manual work" (*Ennead* III.8; Armstrong 1967, 373). Moreover, unable to contemplate ideas in their totality, Soul has, as it were, to move from one idea to another. Thus, the "life of soul is the life of discursive reason in which soul presents one activity after another" (Smith 1996, 210).

Individual souls are connected to the *hypostasis* Soul and have a similarly complex structure. Just as the latter, they have the potential to contemplate ideas and, ideally, even reach the experience of unity with the One in the process called "ascension" (Bussanich 1996, 38-42). But, at the same time, they are also exposed to the world of senses and all the potential distractions, desires and temptations coming from it (Clark 1996, 289). In fact, matter, the material constituent of physical reality, is viewed by Plotnius as "evil in itself" because it is deprived of being and intelligible ideas. Body is therefore only a "shadow of being" and a "second evil" (Corrigan 1996, 107-108).

Contemplative Return

Plotinus often refers to the emanation process beginning with the One and ending at the level of individual souls and their experiences as descent, which at first glance seems to indicate a linear downward movement. However, a closer look on this series of generations shows a rather complex picture, in which "turning back" (*epistrephein*) plays a central role. The following passage from *Ennead* V.2 seems to be the most illuminating in this respect, also highlighting the nature of the causal link between the *hypostasis.*

> This, we may say, is the first act of generation: the One, perfect because it seeks nothing, has nothing, and needs nothing, overflows, as it were, and its superabundance makes something other than itself. This, when it has come into being, turns back upon the One and is filled, and becomes Intellect by looking towards it. Its halt and turning towards the One constitutes being, its gaze upon the One, Intellect. Since it halts and turns towards the One that it may see, it becomes at ones Intellect and being.

> Resembling the One thus, Intellect produces in the same way, pouring forth a multiple power—this is a likeness of it—just as that which was before it poured it forth. This activity springing from the substance of Intellect is Soul, which comes to be this while Intellect abides unchanged: for Intellect too comes into being while that which is before abides unchanged. But Soul does not abide unchanged when it produces: it is moved and so brings forth an image. It looks to its source and is filled, and going forth to another opposed movement generates its own image, which is sensation and the principle of growth in plants. (Armstrong 1984, 59-61)

Accordingly, Intellect is actualized only after its incipient form, which comes about following the "overflowing" of the One, "turns back" to and "gazes upon" it.[24] Soul is also reported to originate from Intellect's "pouring forth" and is able to "look to its source", implying the same movement of turning back. Therefore, this (contemplative) return and not the "overflowing" is sometimes regarded as the "determining element" in the metaphysics of Plotinus (Gatti 1996, 31). Although "overflowing" and "turning back" seem to imply movement, Plotinus is adamant that the One and Intellect are completely motionless, and that movement enters the picture only at the level of Soul, which "does not abide unchanged" and "is moved."

Eternity and Time

In discussing eternity, Plotinus focuses on the first two elements of the intelligible reality, the One and—even more explicitly—Intellect. While eternity is not identical with Intellect, it is "something which, as it were, shines out from the substrate" (Armstrong 1967, 305). As such, it is closely connected with some of Intellect's features, preeminently unity and rest. The all-encompassing world of Intellect is characterized by the identity of thinker and thought, a non-discursive *noesis* of itself. Therefore, any movement is inconceivable for it. These features constitute eternity:

---

[24] In *Ennead* V.1 Plotinus seems to be claiming that Intellect is produced as the result of the One returning to and reflecting on itself (Beierwaltes 1996, 14-16). Armstrong is opposed to this view, for "there can be absolutely no separation from itself or multiplicity in the One" (Armstrong 1984, 34n1). Instead, the subject of the sentence in question should be Intellect. Regardless of which view is correct, this passage offers strong evidence for the importance of return in Plotinus' metaphysics.

> The life, then, which belongs to that which exists and is in being, all together and full, completely without extension or interval, is that which we are looking for, eternity. (Armstrong 1967, 305)

In some passages, he further identifies the intellectual contemplation of the One as a constitutive feature of eternity:

> For that which is this and abides like this and abides what it is, an activity of life abiding of itself directed to the One and in the One, with no falsehood in its being or its life, this would possess the reality of eternity. (1967, 313)

At the same time, eternity is not to be equated with a long temporal interval, since the latter already presupposes the existence of time (Smith 1996, 203). Instead, eternity has to be thought of as a timeless or atemporal state of existence.

Given the overall metaphysical picture, which stipulates that only Soul among the intelligible entities is capable of movement, it comes as no surprise that Plotinus makes the latter responsible for the emergence of time. The reason for time's appearance is described as follows:

> There was a restlessly active nature which wanted to control itself and be on its own, and chose to seek for more than its present state, this moved, and time moved with it; and so, always moving on to the "next" and the "after," and what is not the same, but one thing after another, we made a long stretch of our journey and constructed time as an image of eternity. (1967, 339)

Accordingly, Soul's "restlessly active nature" and the desire to "seek for more than its present state" caused it to abandon the eternal state of self-observation and enter into time, that is, into the intelligible realm where ideas are not observed at once but in temporal succession. In other words, time is caused by the discursive thinking of Soul (Smith 1996, 210).

Because Soul imbues the physical reality with temporality, time exists on two levels: as the life of Soul and as perceived in the physical world (Smith 1996, 210).[25] However, because the emergence of natural existence follows from Soul's internal structure, which is timeless, the generation of the physical world itself is not a temporal process. Soul does not "take

---

[25] Plotinus' main criticism of the traditional theories of time was their attempt to explain time through the motion of bodies in the physical reality. In his account, these physical motions presuppose and take place in time, which goes back to a higher level of reality, Soul, and its "discursive reason."

time" in creating the physical reality (Wagner 2008, 281). In this way, the entire "descent" from the One to sensory reality is a timeless happening.

To resolve the obvious inconsistency that Plotinus views some movements of intelligible entities as atemporal (overflowing, turning back, descending) and some as constitutive of time (movements to the next and after), Michael Wagner distinguishes between the vertical and horizontal movements of Soul. Accordingly, while the vertical movement is characteristic of all intelligible entities, the horizontal component constitutes a unique feature of Soul, "generatively responsible for past and future in regards to natural and corporeal existence" (2008, 358).

The return or ascension of an individual soul to the intelligible reality of Intellect and even the One is a main topic in Plotinus' work. In fact, it might be a reflection of the philosopher's own mystical experiences. Porphyry reports that during the years of his studies with Plotinus, the latter experienced union with the One no less than four times (Reale 1990, 306). Since the One and Intellect are eternal realities, the successful ascension of an individual soul has the effect of leaving time and entering eternity: "when soul . . . returns to unity time is abolished" (Armstrong 1967, 345). Because eternity is a timeless state which can only be realized in the intelligible realm, there can be no corresponding manifestation to it in the physical world. To return to Plotinus' biography, despite his numerous mystical experiences, the span of his life measured 65 years which was not very long, even by the standards of ancient Rome.

Conclusion

The above analysis has shown several similarities and differences between the conception of time in the *Laozi* and the *Enneads* ascribed to Plotinus.

With regard to similarities, it has become clear that the two envisioned time to come into play first at the prenatural stage of the cosmogonic process, that is, before the emergence of physical reality. However, while in the *Laozi* the most likely elements to produce time are entities associated with numbers one, two, and three, Plotinus identifies Soul as the locus of time emergence.

Furthermore, in their temporal deliberations the two works utilize metaphors associated with both linear and cyclical motions and spatial characteristics. On the one hand, this goes against the persistent belief that the Chinese time model was predominantly cyclical (e. g., Pankenier 2004, 130-31); on the other hand, it supports recent comparative investigations into time metaphors in Mandarin and English, which suggest that the only difference between them "lies not in their conceptualization but in the relative frequency of their use" (Link 2013, 137).

With regard to differences, the main item is their view of the temporal characteristics of the highest reality. It appears that the *Laozi* was not particularly interested in associating the Way with timelessness, as following from its "empty" characteristics. Instead, the text formulates the principle according to which the Way can endure under conditions of incessant change. This principle is return and the perpetuity achieved through it is in fact a form of continuous duration. The reason for this may have been the "holistic" worldview of the early Chinese, who did not separate a higher reality and the human world (Allan 1997, 22). Accordingly, even the highest principles had practical significance for humans as objects of emulation. While many of the formulations of the One in Plotinus can also be applied to the Way, their temporal significance is very different. The One in the *Enneads* is eternal, but it does not endure, because its eternity is ultimately founded in timelessness.

Lastly, whereas in the *Laoizi* return is mainly associated with the downward movement, Plotinus views it as the opposing ascending motion. Moreover, in the former, return can be accomplished in many different areas, such as cosmology, politics, ethics, personal cultivation etc. In most of these cases, personal cognitive grasp of the Way is not necessarily part of the process. For Plotinus, on the other hand, return can be completed only as a noetic vision of the Intellect's ideas. Even though certain moral virtues are recognized as conducive to the success of this intellectual endeavor. This shows that the scope of Laozi's return is much wider than it was conceived by Plotinus.

Bibliography

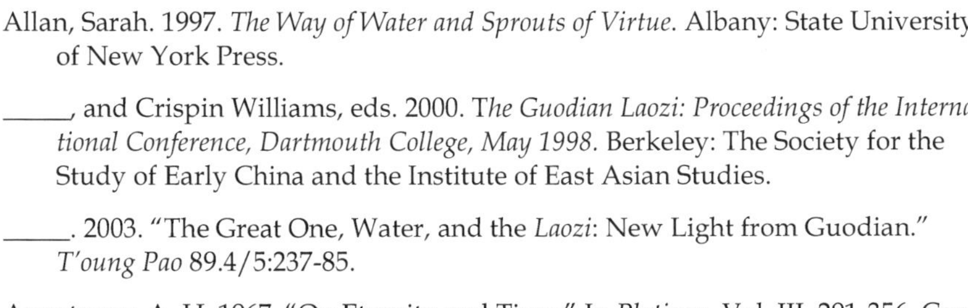

Allan, Sarah. 1997. *The Way of Water and Sprouts of Virtue*. Albany: State University of New York Press.

_____, and Crispin Williams, eds. 2000. *The Guodian Laozi: Proceedings of the International Conference, Dartmouth College, May 1998*. Berkeley: The Society for the Study of Early China and the Institute of East Asian Studies.

_____. 2003. "The Great One, Water, and the *Laozi*: New Light from Guodian." *T'oung Pao* 89.4/5:237-85.

Armstrong, A. H. 1967. "On Eternity and Time." In *Plotinus*, Vol. III, 291-356. Cambridge, Mass.: Harvard University Press.

_____. 1984. *Plotinus*, Vol. V. Cambridge, Mass.: Harvard University Press.

_____. 1988. *Plotinus*, Vol. VI. Cambridge, Mass.: Harvard University Press.

Beierwaltes, Werner. 1995. "Einleitung." In *Über Ewigkeit und Zeit*, 9-88. Frankfurt am Main: Klostermann.

Bokenkamp, Stephen R. 1999. *Early Daoist Scriptures*. Berkeley: University of California Press.

Bussanich, John. "Plotinus' Metaphysics of the One." In *The Cambridge Companion to Plotinus*, edited by Lloyd P. Gerson, 38-65. Cambridge: Cambridge University Press.

Chan, Allan K. L. 1991. *Two Visions of the Way: A Study of the Wang Pi and the Ho-Shang Kung Commentaries on the Lao-Tzu*. Albany: State University of New York Press.

Chang, Leo S., and Yu Feng. 1998. *The Four Political Treatises of the Yellow Emperor*. Honolulu: University of Hawai'i Press.

Chen Guying 陳鼓應. 2006. *Laozi jinzhu jinyi* 老子今注今譯. Beijing: Shangwu yinshua guan.

Chen Ligui 陳麗桂. 2017. "Dao de yicheng ji qi yihan yanhua" 道的異稱及其意涵衍化. In *Gujian xinzhi: Xi Han zhushu Laozi yu Daojia sixiang yanjiu* 古簡新知：西漢竹書老子與道家思想研究, edited by Han Wei 韓巍, 315-45. Shanghai: Shanghai guji chubanshe.

Chen Mengjia 陳夢家. 2016. *Laozi fen shi* 老子分釋. Beijing: Zhonghua shuju.

Cheng, Chung-ying. 1974. "Greek and Chinese Views on Time and the Timeless." *Philosophy East and West* 24.2:155-59.

Clark, Stephen R. L. 1996. "Plotinus: Body and Soul." In *The Cambridge Companion to Plotinus*, edited by Lloyd P. Gerson, 275-91. Cambridge: Cambridge University Press.

Cook, Scott. 2012. *The Bamboo Texts of Guodian: A Study and Complete Translation*. Ithaca: Cornell University.

Corrigan, Kevin. 1996. "Essence and Existence in the *Enneads*." In *The Cambridge Companion to Plotinus*, edited by Lloyd P. Gerson, 105-29. Cambridge: Cambridge University Press.

Dy, Manuel B., Jr. 1996. "The Chinese View of Time: A Passage to Eternity." In *The Humanization of Technology and Chinese Culture*, edited by Tomonobu Imamichi, Wang Miaoyang, and Liu Fangtong, 251-68. Washington, DC: Council for Research in Values and Philosophy.

Erkes, Eduard. 1945. "Ho-Shang-Kung's Commentary on Lao-tse." *Artibus Asiae* 8. 2-4:121-96.

Fech, Andrej. 2015. "The Protagonists of the Daoist Treatise *Wenzi* in Light of Newly Discovered Materials." *Oriens Extremus* 54:209-48.

_____. 2018. "Reflections on Artisan Metaphors in the *Laozi*: Who Cuts the "Uncarved Wood" (*pu* 樸)? (Part 1)." *Philosophy Compass* 13.3.

Gatti, Maria Luisa. 1996. "Plotinus: The Platonic Tradition and the Foundation of Neoplatonism." In *The Cambridge Companion to Plotinus*, edited by Lloyd P. Gerson, 10-37. Cambridge: Cambridge University Press.

Gerson, Lloyd P. 1996. "Introduction." In *The Cambridge Companion to Plotinus*, edited by Lloyd P. Gerson, 1-9. Cambridge: Cambridge University Press.

Girardot, N. J. 1988. *Myth and Meaning in Early Taoism*. Berkeley, Los Angeles, London: University of California Press.

Graham, A. C. 1998. "The Origins of the Legend of Lao Tan." In *Lao-tzu and Tao-te-ching*, edited by Livia Kohn and Michael Lafargue, 23-40. Albany: State University of New York Press.

_____. 1989. *Disputers of the Tao: Philosophical Argument in Ancient China*. La Salle: Open Court Publishing Company.

Jhou, Nihel. 2020. "Daoist Conception of Time: Is Time Merely a Mental Construction?" *Dao* 19.4:583-99.

Hall, David. L., and Roger T. Ames. 1995. *Anticipating China: Thinking Through the Narratives of Chinese and Western Culture*. Albany: State University of New York Press.

Harbsmeier, Christoph. 1995. "Some Notions of Time and of History in China and in the West." In *Time and Space in Chinese Culture*, edited by Chun-Chieh Huang and Erik Zürcher, 49-71. Leiden: Brill.

He Ning 何寧, ed. 1998. *Huainanzi jishi* 淮南子集釋. Beijing: Zhonghua shuju.

Henricks, Robert G. 1989. *Lao Tzu: Te-Tao Ching – A New Translation Based on the Recently Discovered Ma-wang-tui Texts*. New York: Ballantine Books.

_____. 2000. *Lao tzu's Tao Te Ching: A Translation of the Startling New Documents Found at Guodian*. New York: Columbia University Press.

Kaltenmark, Max. 1968. *Lao Tzu and Taoism*. Translated from the French by Roger Greaves. Stanford: Stanford University Press.

Kenny, Anthony. 2004. *Ancient Philosophy: A New History of Western Philosophy*, Vol. 1. New York: Oxford University Press.

Kim, Hongkyung. 2012. *The Old Master: A Syncretic Reading of the Laozi from the Mawangdui Text A Onward*. Albany: State University of New York Press.

Klein, Esther S. 2013. "Constancy and the Changes: A Comparative Reading of *Heng Xian*." *Dao: A Journal of Comparative Philosophy* 12:207-24.

Kohn, Livia. 1998. "The Lao-tzu Myth." In *Lao-tzu and Tao-te-ching*, edited by Livia Kohn and Michael LaFargue, 41-62. Albany: State University of New York Press.

Lau D. C. 2001 [1963]. *Tao Te Ching*. Hong Kong: The Chinese University Press.

Lewis, Mark Edward. 2006. *The Flood Myths of Early China*. Albany: State University of New York Press.

Link, Perry. 2013. *An Anatomy of Chinese. Rhythm, Metaphor, Politics*. Cambridge, Mass.: Harvard University Press.

Liu Xiaogan 劉笑敢. 2016. *Laozi gujin* 老子古今. Beijing: Zhongguo shehui kexue chubanshe.

_____. 2017. "Jianbo ben Laozi de sixiang yu xueshu jiazhi--yi Beida Hanjian wei qiji de xin kaocha" 簡帛本老子的思想與學術價值: 以北大漢簡為契機的新考察. In *Gujian xinzhi* 古簡新知, edited by Han Wei 韓巍, 103-27. Shanghai: Shanghai guji chubanshe.

Loewe, Michael. 1995. "The Cycle of Cathay: Concepts of Time in Han China and Their Problems." In *Time and Space in Chinese Culture*, edited by Chun-Chieh Huang and Erick Zürcher, 305-28. Leiden: Brill.

Nienhauser, William H. Jr., ed. 1995. *The Grand Scribe's Records. Volume I: The Basic Annals of Pre-Han China*. Bloomington: Indiana University Press.

Queen, Sarah A. 2013. "*Han Feizi* and the Old Master: A Comparative Analysis and Translation of *Han Feizi* Chapter 20, "Jie Lao," and Chapter 21, "Yu Lao"." In *Dao Companion to the Philosophy of Han Fei*, edited by Paul R. Goldin, 197-256. New York.: Springer.

Pankenier, David W. 2004. "Temporality and the Fabric of Space-Time in Early Chinese Thought." In *Time and Temporality in the Ancient World*, edited by Ralph M. Rosen, 129-46. Philadelphia: University of Pennsylvania Museum of Archaeology and Anthropology.

Perkins, Franklin. 2015. "*Fanwu Liuxing* 凡物流行, 'All Things Flow into Form,' and The One in the *Laozi*." *Early China* 39:195-232.

Reale, Giovanni. 1990. *A History of Ancient Philosophy IV: The Schools of the Imperial Age*. Translated by John R. Catan. Albany: State University of New York Press.

Seidel, Anna, and Lothar von Falkenhausen. 2008. "The Emperor and His Councilor Laozi and Han Dynasty Taoism." *Cahiers d'Extrême-Asie* 17:125-65.

Smith, Andrew. 1996. "Eternity and Time." In *The Cambridge Companion to Plotinus*, edited by Lloyd P. Gerson, 196-216. Cambridge: Cambridge University Press.

Unschuld, Paul U. 2016. *Huang Di Nei Jing Ling Shu: The Ancient Classic on Needle Therapy*. Berkeley: University of California Press.

Wagner, Michael F. 2008. *The Enigmatic Reality of Time: Aristotle, Plotinus, and Today*. Leiden: Brill.

Wagner, Rudolf G. 2003. *A Chinese Reading of the Daodejijng: Wang Bi's Commentary of the Laozi with the Critical Text and Translation*. Albany: State University of New York Press.

Wang Zhongjiang 王中江. 2015. *Chutu wenxian yu Daojia xinzhi* 出土文獻與道家新知. Beijing: Zhonghua chubanshe.

_____. 2019. "The Concept and Genealogy of the Ultimate Origin: An Exploration of Constancy in the *Hengxian* Text of the Shanghai Museum Collection." *Journal of Chinese Philosophy* 46.1-2:3-32.

_____. 2019. "Abnormalities and Return: An Exploration of the Concept *Fan* 反 in the *Laozi*." *Religions* 10:32; doi:10.3390/rel10010032.

Wohlfart, Günter. 2001. *Der philosophische Daoismus: Philosophische Untersuchungen zu Grundbegriffen und komparative Studien mit besonderer Berücksichtigung des Laozi (Lao-tse)*. Köln: ed. chora.

Xing, Wen. 2015. "Early Daoist Thought in Excavated Bamboo Slips." In *Dao Companion to Daoist Philosophy*, edited by Xiaogan Liu, 101-28. New York: Springer.

Magic Square and Perfect Sphere

Time in Daoism and Ancient Greece

NADA M. SEKULIC

> The Way (Dao) comes into being by people walking.
> Things come into being by people naming. (*Zhuangzi*)

Daoist philosophy has always focused on studying the rhythmic cycles of nature in time and space, believing that the deepest wisdom comes from understanding their ongoing dynamic. Unlike ancient Greek philosophy at the root of all Western thought, it centers more on comprehending the cosmic processes, stages, transitions, and transformations than on establishing distinct lines and differentiating between the more static dimensions of form and essence, senses and mind, chaos and cosmos, *physis* and *nomos*.

In general, moreover, philosophers both in early Greece and ancient China (6th-3rd c. BCE) tend to contain many religious elements, only gradually engaging with scientific evidence, applied knowledge, political debate, and strategic application. The abstract and religious cannot always be clearly divided, and philosophers at the time had to be both erudites and sages, some even being considered demigods.

Comparing the philosophy of the two areas, Xu Keqian (2010) argues that ancient Greek philosophers focused on the exploration of truth while Daoist thinkers concentrated more on the exploration of Dao. Starting from this point, he outlines a number of parallels and differences. However, truth and Dao do not entirely match. In pre-Socratic thought, the word for "truth" was *aletheia*, meaning "disclosure" and indicating something unconcealed. The term may also be translated as "direct insight" into factual reality. The most appropriate match of "truth" in this sense in classical Chinese is *zhen* 真 rather than *dao* 道, indicating that, when truth unveils itself, consciousness and reality are indistinguishable.

It seems more appropriate to compare the term *dao* to the ancient terms *nous* or *logos*, since all three refer to the universal foundation of the cosmos, to the hidden structure, order, and law underlying everything. *Nous* and *logos* indicate unity and compatibility between thought and cosmos; they express the deep belief of the archaic Greeks ever since Parmenides that rational thought and reality must coincide, be the same, relate to the same, or have a corresponding relationship—a principle fully and with great precision formulated later by Aristotle. Therefore, in ancient Greece, the practice of philosophical self-cultivation was primarily linked to logic and based on reason and dialectic or dialogue (Heidegger 1984, 5-34).

This perspective of the universe implied the idea of two planes of existence—one apparent and one hidden—but still existing, like an invisible root, which is not self-evident. The apparent is related to the phenomenal world of opinion, and the hidden one with the unchanging and absolute basis of the universe that can only be reached by logic and thought. Time in that division belongs to illusion—it is a measure of change and has meaning only for those who experience it. For example, the time of the great celestial cycles (the relative positions of the fixed stars and the planets) shape a different time than that of humanity, which is insignificant and almost non-existent in relation to them. The essence of time is determined by the interrelationship of perishable things and processes.

For the ancient Greeks until Plato, time has no beginning and no end, but appears in cycles, large and small, and is therefore like a seed from which everything is born and disappears again. What is revealed behind it (*meta*) contains the logos, the law of events expressed by the nature of logical thinking, as that which can be known in advance about the things of the universe (*tà mathémata*).

For ancient Daoist thought as presented in the *Zhuangzi*, change occurs beyond prediction and this is an inherent property of the universe. It can be followed, but not comprehended or controlled. That is why the process of coming to fundamental truths is connected with the task of emptying the mind and following the natural process or becoming of things. In this process time and its nature are revealed, that is, time as something that is always at least partly unpredictable and in a constant process of creation, not as something that belongs to the world of illusions.

Consequently, in the Daoist approach, the connection between thought, truth, and reality is far more enigmatic, vaguer, and a great deal more unpredictable. It always relates to cultivation practice and a particular way of living in the world. In Daoism, the human being should align with Dao in order to achieve a state of oneness or some kind of insight that goes beyond words. Reaching this level or vision, attaining the "landscape" of Dao as a whole, is only possible through intuition and by nurturing a certain way of life (Schwartz 1985). It is a dynamic process, a going-along with the rhythmic, ever-changing patterns of nature rather than a stabilizing, static inquiry, a sophisticated logical reflection, as preferred by the ancient Greeks.

Let us say, somewhat simplified and in modern language, that the Daoist understanding of time is closer to quantum theory than ancient Greek thought, which requires a consistent and deductively derived logical description of the distinction between illusion and truth. This is so because in Daoist thought it is clear that time both is and is not. Its nature is twofold, it is in constant flux of becoming and disappearing. However, this does not make it illusory and/or distant from the truth, but only obscure to human beings.

In this sense, in ancient Daoist thought, the science of being as a being (being identical to oneself) or metaphysics in the true sense of the word is not possible, nor can the nature and essence of time be explained from this principle. Daoist philosophy requires the notion of becoming as something that is not logically grounded. The usual interpretation of ancient Chinese philosophy sees it as primitive or undeveloped-naturalistic, a notion that goes back to Hegel and his limited understanding of Chinese philosophy, aesthetics, and history (Sekulic 2015). In contrast, Daoists deconstructed metaphysics and metaphysical language, which is the core of Western intellectual tradition, even at the dawn of Chinese intellectual history and can be quoted as an ancient critique of logocentrism (Burik 2006; Sekulic 1997; 2004).

Dao

Dao, both as term and concept, is crucial for understanding Daoist thought. The polysemic and long-term use of the word make it rather intricate and obscure it to a certain degree. The character first appears during the Zhou dynasty (1122-221 BCE) on numerous bronze objects in an archaic bronze script version, on records preserved on bamboo slips, as well as inscriptions marking major rituals and important events. It consists of the graphs for "head," a stylized "eye," and the action of "walking," thus fundamentally indicating intentional, conscious movement.

From here, the character developed multiple meanings that partially overlap, including those used in everyday life as well as in highly sophisticated philosophical discourse. In modern Chinese, its meaning is quite pragmatic, signifying path, street, or direction of movement. However, considering its ancient and philosophical meaning, Dao indicates the highest harmony between humanity and the cosmos, the general underlying and invisible ontological principle, as well as the method and purpose of self-cultivation. In its earliest use, it also indicated a well-governed state, that is, a state established on the principles of Dao, which, according to some authors, even factually existed in the area of present-day Hunan (Boodberg 1957, 598-618).

In terms of cosmology, Dao from an early age was connected to political systems, since rulership had to be established on divine and cosmic principles. Each kingdom and dynasty struggled to prove its divine origin in its own way, and Dao in this context was a prominent feature, despite the fact that the ideal of the Daoist sage did not necessarily imply an active political life (Wang 2006).

In Daoist thought, Dao represents the principle of flowing, growing, and always changing reality. It appears more as a vague symbol than a notion or concept with a clearly defined scope and content. In fact, according to the *Daode jing*, Dao is nameless and ineffable, so that the word presents merely a limited designation for something that cannot be restricted to language. Rather, it represents the "way" as the core of reality, the truly existing principle of maintaining and nurturing the universe. It implies the overarching order inherent in all beings, including humans. They can understand and follow it only through a process of purposive, meaningful, and harmonious living, always maintaining the awareness of social and natural environment. As the *Zhuangzi* says, "Dao comes into being by people walking."

While the ancient Greeks studied change primarily in terms of place, that is, as quantitative, the term for "change" or "becoming" (*hua* 化), so important in Daoist thought, required a qualitative change. Even in modern Chinese, it still forms an element in a series of composite terms, all closely associated with the idea of transformation. Examples include "change" (*bianhua* 變化), "chemistry" (*huaxue* 化学), "purifier" (*jinghuaqi* 净化器), "culture" (*wenhua* 文化), "digestion" (*xiaohua* 消化), and "modernization" (*xiandaihua* 现代化). All these imply that numerous dimensions of life are predicated on change and that time, as its key paradigm, relates to qualitative transformation.

So, when Zhuangzi says that Dao comes into being by people walking, he does not mean "changing place," but undergoing a transformation, leaving the realm of things and names and embracing processes. Time has a relationship to relative terms. Time is not a thing, it is constantly being created, always partly unpredictable, which corresponds to the phenomenal perception of time, as opposed to the idea of time as an illusion, which requires abstraction as a medium for understanding the phenomenal world.

Nonbeing

A key Daoist concept that well demonstrates the difference between ancient Greek and Daoist ways of thinking about time in the greater cosmological context is nonbeing (*wu* 無), a latent, flowing underbelly of the existing world. It finds prominent expression in effortless action or "nonaction" (*wuwei* 無為), a way of moving along with things in flow—unplanned and unpredictable, open to changes and ongoing transformation. Its opposite is being (*you* 有), visible and tangible existence as we know it, which connects to "having action" (*youwei* 有為), that is, an intentional, analytical, deliberate plan of doing associated with specific objects. The latter implies an order that must be constructed and actively maintained, planned and executed in time, establishing a matrix of control over the present that involves conscious study and awareness of the past and careful planning of the future.

Being, as much as "having action," matches the principle that has shaped the basic matrix of Western thinking, while nonbeing and nonaction express the logic of organic, continuously emerging structures not composed of separated parts built one by one. Biological structures are always interacting with and dependent on the environment: their growth is spontaneous and not necessarily related to human purposes. Joseph Needham, a major expert on Chinese civilization in comparison to the West, speaks of an "organismic philosophy of China" (1954).

Grammatically the term "nonaction" consists of two words, "nonbeing," which signifies absence, non-existence, and latency, plus "action" which means "doing." Essentially, it means "acting in nonbeing." In contrast to inactivity, according to an explanation given in the *Zhuangzi*, it represents a way of going with the flow, of living in close familiarity with the natural cycles and their inherent rhythms. It indicates the use of natural patterns of timing and flowing in people's actions, matching nature as it acts in its own unique way. That is to say, if people act in harmony with nature, relating to the rhythms and time patterns of the sun and the moon, a significant part of activities necessary to achieve certain goals become part of a temporal flow (Mair 1998).

Expressed by the principles of spontaneity and emptiness, nonaction marks how the world was created, emerging organically through constant change and transformation from nonbeing as the underlying ground with no demiurge or original creator. Thus, when Laozi compares Dao with the mother of all, he does not introduce a theistic notion but rather emphasizes the biogeneric principle. When he replaces the concept of heaven, dominant in Zhou thought as the potent force underlying and directing all existence, with the more amorphous Dao, he relinquishes the very meaning of intentionality, the ultimate intention of any higher force. Everything that exists moves in temporal cycles of spontaneous emergence and dissolution, making nonbeing and latency a constitutive aspect of reality.

Being

This notion is entirely alien to pre-Socratic thinkers whose central focus is on being. Even Democritus, who accepted the existence of nonbeing, thought about it only in terms of "empty space" while still thinking of atoms, the essential substances of universe, as unchangeable. The same contrast also applies to Western mystical teachings, which are deeply eschatological and mostly theistic. Daoist mysticism aims at recognizing the moment when nonbeing connects to being in everyday life, in everything that surrounds people, in the world of "there is" (Schwartz 1985, 199). The world of transient things is neither less real nor does it represent a degradation of spiritual principles, as it would in Western thought.

Until Plato and in particular his dialogue *Timaeus*, chaos and cosmos, mythos and logos, image and abstraction were mutually interchangeable. From then on, however, the philosophical interpretation of the ancient Greek heritage recognized only rational thinking as generative and elevated it to the most important principle of ancient Greek thought.

According to the Eleatic school of philosophy, whose teachings directly precede Plato's philosophy and which exerted a great influence on all later schools, only the One exists. It is static and without motion, the universal principle of reality that overshadows all, so that that plurality does not really exist and spatial motion as well as temporal flow are an illusion. They proved this with the method of logical argument, best known for Zeno's *aporias*. It is based on discovering the logical contradictions in common sense perception, applying the reduction of the common beliefs to a "paradox," from *para* meaning "behind" and *doxa* indicating "opinion." In this manner, thinkers were able to deny the initial thesis through *reductio ad absurdum* or demonstration to the point where it becomes impossible. The method works with the extraction of truth by apagogical arguments or the appeal to extremes, favoring abstraction over sensory reality. Zeno's arguments against motion and temporal flow are well-known, such as the story of the race of Achilles and the tortoise.

His first argument supposes that one needs to cross a path. In order to achieve that, it is necessary—according to the Pythagorean thesis—to go through an infinite number of points in finite time. However, it is impossible to cross an infinite number of points in finite time, and thus the path constitutes an infinite distance. In fact, no object can cross any distance, because the same difficulty always appears, so we must conclude, therefore, that no movement is possible.

In his second argument, Zeno presents a scene where Achilles and a turtle run a race. Achilles gives the turtle a certain advantage, but by the time he reaches the place where the turtle started, she has already moved on to the next point. When he gets to that point, she has already covered a new distance, no matter how small. Thus, Achilles constantly approaches the turtle, but he can really never reach it due to the Pythagorean assumption that length is composed of an infinite number of points. Therefore, although the Pythagoreans claim the possibility of movement, they themselves make it impossible by their own teaching.

In the third argument against movement, Zeno argues with the Pythagorean understanding that an arrow in flight should occupy a certain position in space, since a certain position in space means to be at rest. He says, "An arrow that looks like it is flying does not actually fly, but stands motionless, because if it were to fly, it would have to be at all times and not be in one place."

Even the ancient Greeks undertook many attempts to solve Zeno's aporia. Aristotle, for example, put forward the philosophical argument that space and time are only potentially divisible in infinity, but are actually finite and continuous. According to him, infinity does not have existence, only potentiality. So, when we talk about time, it is potentially infinitely divisible, but actually it is a continuum and its punctuated nature relates to "momentum."

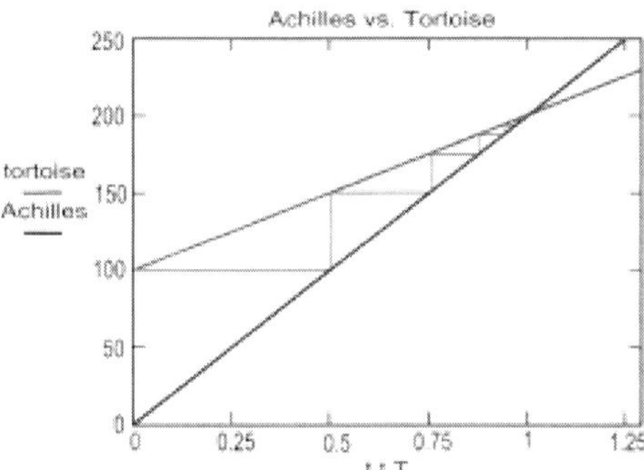

Archimedes, as shown in this chart, derived mathematical proof through geometric convergent sequences. To him "Achilles will catch up with the turtle" because in each step the time it takes to catch up with the turtle is less and less than the time the turtle is moving ahead with each step. His algebraic solution to this problem is also based on deriving finite convergent values in determining time and space.

In Comparison

Ever since the Eleatic school, logical experiments and proof arguments have been part of Western philosophy. Any explanation of the cosmos has relied exclusively on the power of logical thinking. A key question was: How can something moving at finite speed in finite time cross infinitely many discrete points?

Early thinkers found the solution in the sharp separation between appearance and reality, the rational and the irrational, logos and mimesis, chaos and cosmos, knowledge and opinion, potential and actual reality, being and nonbeing. From this eventually came the separation of body and mind, matter and spirit. Not yet fully formulated in Plato or even Aristotle, it has remained essential to the Western identification of philosophy and philosophical self-cultivation with only one side of these polarities (Nietzsche 1977)

In Chinese philosophy, and especially in the Daoist tradition, this division never happened, partly because of the nature of the Chinese language with its pictorial writing system and inherently dynamic, music-like grammatical structures (Kohn 2021, ch. 1). Here images form an important element of meaning and dynamics are at the very root of thinking. Chaos and cosmos, being and nonbeing, myth and reason closely intertwine. Rather than mutually exclusive or linearly related, they complement each other and interact in dynamic flow.

A yet different dimension of the contrast between nonbeing and being is the central focus on society versus cosmos. Ancient Greek philosophy, and Western civilization in its wake, concentrated strongly on the rational study of world, focusing on problems of human society as expressed in the notion of the state (*polis*), issues of rational thinking, and by codifying education in academia.

Daoists, on the other hand, maintained a dominant focus on the cosmos, developing the ideal of immortality (*xian* 仙/仚). Immortals are people whose path, way of thinking, and modes of life are different from ordinary society, linked in in harmony with nature and moving in close connection to the cosmos. According to the *Zhuangzi*, they live in complete spontaneity and keep away from politics and social obligations. Accepting death as a natural transformation, they show appreciation and praise for things that others may view as useless or aimless and stridently reject standard social values and conventional reasoning.

The word for "immortal" consists of the graphs for "person" and "mountain," the latter being more than a mere symbol. Mountains are sacred places, and immortals are hermits, people who have retreated into nature, gone into the mountains to grasp the inner meaning and order of things and avoid the decline and decay that come from leading a worldly life. Daoist philosophy is clear on this point: mortal people can transform themselves into subtler levels of being, into vital energy (*qi* 氣) and pure spirit (*shen* 神), and thereby become immortal.

Here the realm of the gods and the world of human beings are not insurmountably separate and their temporalities remain linked. Transformation and holiness are immanent to the earthly world, a feature also expressed in the strong promotion of physical well-being, health enhancement, and longevity—attained through the practice of a variety of techniques that lead through healing and long life to the transformation into an immortal. Time in this vision is one system: human time remains embedded in cosmic time, the life expectancy of anyone can be extended indefinitely, the spirit at the root of the cosmos is infinite, and human beings partake in eternity at all times.

Greek philosophy, in contrast, gradually relinquished the close connection to the gods, and immortality became the totally other, reflecting a temporality that may be unlimited but has nothing in common with human time, which is always structured and always contained. With the disappearance of the concept of *arche*, the seed of the world, characteristic of pre-Socratic thought, it lost its organic character. Ancient sages here cherished autarky (*autarkeia*) as a contemplative attitude towards life, but this has nothing to do with nature or the cosmos. Philosophers are not linked to mountains: they are part of the *polis* and the academia— their bodies limited, their lives contained, their time short.

The *Luoshu*

Daoists express the ongoing flow and infinite transformation of life in a variety of visual and numerical symbols. The best known among them is the pictorial expression of the interdependent forces of yin and yang, the energetic dynamic of the universe, known as the Diagram of the Supreme Ultimate (*Taiji tu* 太極圖). This shows the unity of opposites in the cosmic processes, the never-ceasing flow of change, the potent structure of all reality. While in its best-known form only found since the 17th century (Louis 2003), it appears in a more rudimentary version in a chart attributed to the Neo-Confucian thinker Zhou Dunyi 周敦義 (1017-1073) (Wang 2005).

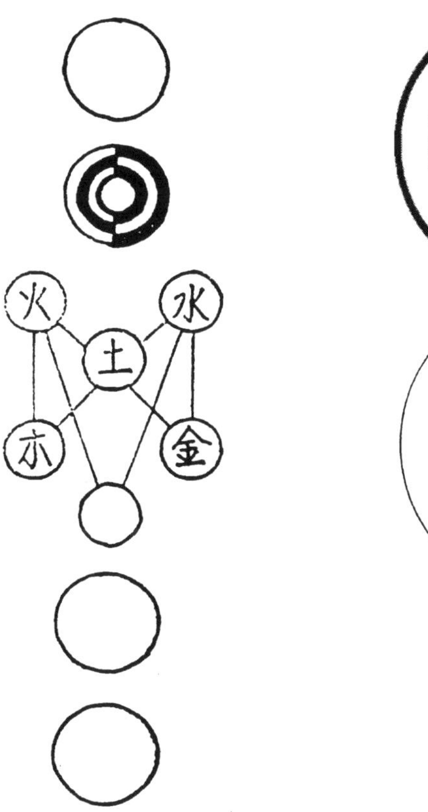

Zhou Dunyi's *Taiji tu*

Diagrams of Yin and Yang

Historically, moreover, the yin-yang diagram goes back to ancient magic squares, a potent way of visually translating the decimal into a binary numerical system. Diagrams and charts as much as ancient talismans were dynamic visual symbols or pictorial metaphors that contained cosmic potency and activated certain proportions or different dynamics of reality in unique constellations, closely related to personal destinies as well as collective events. They were used in divination, for medical purposes, as seals, and in rulers' insignia, inextricably linked to the development of the Chinese script, in itself a representation of the magical, prophetic, and ritualistic dimensions of life (Verellen 2006).

The most famous among all the various magic squares is the *Luoshu* 洛書 (Writ of the Luo [River]), allegedly transmitted to the sage-king Yu—famous for taming the all-consuming flood—by a sacred turtle rising from the river. The turtle, with its physical structure of flat underbelly and round shell, in traditional China was seen as a representation of heaven and earth (see Allan 1991), while its extended life span made it a symbol of wisdom, longevity, and immortality.

Another origin tale also connects the *Luoshu* to the great flood, but in a different manner. To stop the waters, people offered various sacrifices to the river god, but he would not accept them. Rather, each time a turtle would rise from the waters, and the deity only accepted the sacrifice when a child noticed the markings on the turtle's shell. They were:

8	1	6
3	5	7
4	9	2

In this square of three-times-three sections, odd numbers stand for yang and heaven, while even numbers represent yin and earth. They alternate, much like the yin-yang diagram symbolizing the continuous change of opposites and showing the ratio of binary pairs. The four even numbers appear in the far corners of the square, while the five odd numbers form a cross, with number five in the middle. They always add up to fifteen, a core number of the Chinese calendar, which divides the year into twenty-four solar periods of fifteen days each.

Thus, the link of the square to annual floods, divine animals, and ritual sacrifices in the origin stories relates to pertinent knowledge of weather, calendar, and seasonal change as well as the exploration of mathematics as magic and magic as mathematics. The number five in the central cell, then, connects each pair of numbers horizontally, vertically, or diagonally whose basic sum equals ten, with five marking the central pivot. In addition, each corner represents a geographical direction (north, northeast, east, southeast, etc.)—nine always being south and one standing for the north.

Taken together, the *Luoshu* represents a complex mandala of spatial and temporal proportions in the natural processes of the universe. It allows extensions and permutations of numbers with repeated regularities

(Swetz 1979). Beyond its obvious ritual and magical purposes, it still is important for the organization of space according to Fengshui and has great relevance in applied mathematics. The well-known Pythagorean theorem regarding the relation of the sides of a right-angled triangle, too, is a magic square based on the number three. Known in China even earlier, it was preceded by an algebraic and arithmetic exploration of Pythagorean triples, developed in Greece into greater sophistication and mostly theoretical proofs of geometry.

The oldest Chinese text with formal mathematical explanations is the *Zhou bisuan jing* 周髀算經 (Zhou Classic of Measuring and Calculation). It shows that Zhou-dynasty mathematicians knew of the principles of empirical geometry and made use of what we call the Pythagorean theorem, providing the oldest known demonstration of its validity (Kline 1962). Another key feature of ancient Chinese mathematics is their development of the binary number system, prominently applied in the *Yijing*, which the 17th-century thinker Gottfried Wilhelm Leibniz (1646-1716) interpreted as an early form of binary calculus. He noted that the hexagrams corresponded to binary numbers from 0 to 111111, and concluded that the *Yijing* proved a major Chinese accomplishment in philosophical mathematics (Strickland 2006).

To sum up, the *Luoshu* forms an integral part of the mathematical inventions of ancient China, which also include the accurate determination of Pi, the use of zero, the application of decimal and negative numbers, the method of extracting square roots, and the solution of equations with multiple variables. Ancient Chinese mathematicians solved geometric problems with algebra rather than rulers and calipers as typical for Plato's academy with its dominant focus on geometry, also typical for mathematics in Egypt and Mesopotamia. In China, on the other hand, geometry was secondary to, and evolved from, algebra (Temple 1986)—numbers relating to the dynamics of time and space where geometrical forms connect to static and stable elements.

The Five Phases

Another dimension of the magic square in ancient China is its central focus on the number five, which links it to the system of the five phases (*wuxíng* 五行). Marking the dynamic unfolding of yin and yang through time and space, they signify the stages of lesser yang, greater yang, yin-yang, lesser yin, and greater yin, and are symbolized by natural features that relate to each other dynamically: wood, fire, earth, metal, and water (see Martzlow 2016).

Change is the very essence of these phases, and the word *xing* literally means to progress, walk, or move forward. Rather than existing one by one and denoting a given property or matter, they are markers of relationships, of dynamic connections. Reality in this system is a "bricolage," to use a term coined by Claude Levi-Strauss, something created from whatever is available nearby, rather than an emanation of stable, solid underlying perfection. It is more repetition that can be calculated than an expression of an immutable law. This also matches the nature of mathematics: in ancient China they were primarily inductive and focused on discovering regularities while in Greece they were deductive, oriented towards proof and focused on theoretical issues.

The five phases soon came to epitomize the transformation of the political order, applied sporadically and more cosmologically at first, then forming the backbone of all correlative thinking at the root of imperial and social reality (see Graham 1986). A product of the highly unstable political situation during the Warring States and associated with turbulent changes and wars (Wang 2006, 5), they yet also exemplify the central focus on dynamic process thinking in China, the complete absence of any idea of a static, perfect ground. Change is the basis of everything, so that both being and nonbeing are inherent attributes of reality. Magic squares accordingly show the power of change rather than the perfection of form.

The same pattern also applies to Chinese medicine, which—unlike in the West, where the body consists of fixed anatomical features—sees the human being as a flowing network of energy conduits centered on the five inner organs that represent the five phases within the person. It also focuses greatly on dynamic bodily functions such as breathing, temperature fluctuations, digestion, tissue nutrition, moisture maintenance, and the like, seeing them as the result of energetic dynamics closely related to circadian and seasonal rhythms.

Diagnostics are based on recognizing the patterns of disharmonious functioning of vital energy, and disorders are ideally detected before a specific illness occurs, since disease is a physical manifestation of an energy disorder induced by external events such as trauma or internally through genetic patterns or emotional disturbance. Thus, the organs are agents of various bodily functions and manifestations of energetic properties rather than specific material centers recognized in Western medicine. Transformation of one phase into another, one quality into the next, is at the core of this system, which moves in temporal patterns and closely relates to natural phenomena. Going far beyond healing and recovery, the five phases in the human body are the core of health enhancement, longevity, and even immortality (Kohn 2006).

The Greek Cosmos

The most complete description of the ancient Greek understanding of the cosmos appears in Plato's *Timaeus*, which outlines the creation of the universe and humanity (1995). It is preceded by his interpretation of Parmenides' teaching, that is, of the Eleatic doctrine, in the dialogue *Parmenides*. According to this, the thinker taught that the characteristics of being are: 1) not generated (*agenêton*) and imperishable (*anôlethron*); 2) whole and unique, i.e., the only thing of its kind (*oulon mounogenes*); and 3) unshaken (*atremes),* that is, absolute now and continuous. His teachings imply that being is one, not many; there is no emptiness in any part of it; and it can be represented by the perfect sphere, a purely abstract form that does not actually exist anywhere in nature.

He argues that if you want to understand what being truly is—the only thing that can be truly known—then you have to exclude all the temporal and spatial differences, motion, and change that we are used to from the perceptible world. That is, one has to get rid of all those notions central to natural philosophy.

Characteristically, in Plato's *Timaeus*, everything originates from the One, which means that his starting point is taken from Parmenides' teaching. All manifestations of the universe are created by dividing the One, which is full and perfect: it contains no emptiness. Whole numbers are all powerful and held sacred. According to tradition, when the Pythagoreans discovered that the sides of a square are incommensurable with its diagonal, which required the introduction of irrational numbers, they kept it a big secret. When a disciple revealed it, he was forced to commit suicide since the very idea of irrational numbers cast doubt on the basic principle of understanding the universe.

Since there can be no emptiness in the universe, time and creation must have a beginning, and worldly time is but an animated image of eternity whose real essence is form. Even the introduction of the concept of *khora*—space, interval, the shapeless shape—also described as "the mother of all things" or the "third factor"—did not change this. Space participates in creation together with form: the demiurge did not leave room for the concept of nonbeing as a constitutive part of the universe.

Rather, the perfect form of the soul of the universe, the animated demiurge, is seen as a perfect sphere, which moves circularly and uniformly around its axis and requires nothing beyond itself. This metaphysical description of the geometric construction of the sphere has been interpreted as the perfection of the form and served as a metaphor of the demiurge. Everything that exists is inscribed in that form. The transition of the universe to a manifest state, which creates different geometric shapes from the original sphere, is determined by the intention of the creator.

The theistic principle, the introduction of a creator, forms a necessary element of this teaching, since forms are static, abstract structures that precede reality, molds into which all motion must be imprinted. Due to this conceptual framework, the importance of change is diminished. It represents all that is perishable and decaying in the cosmos. In other words, rather than continuous unfolding, creation is a craft business. The demiurge as the ultimate artisan brings forth a geometric imprint of himself, manifest in sets of shapes, and infuses life into each of these forms as in a mold. Once put in motion, moreover, they function as independent mechanisms, but only to a certain degree.

The entire manifest universe, then, was created from five basic primordial forms, mathematical shapes or convex regular solids that can transform into one another. Whereas the Chinese model is algebraic and relates to time, the Greek system is geometric and connects to to space. The five elements at their root are fire, water, air, and earth, as well as the cosmos as a whole. Mathematically they correspond to geometrical shapes that have the same-size surface on all sides and come with three, six, eight, twelve, and twenty faces—known as tetrahedron, hexahedron, octahedron, dodecahedron, and icosahedron, respectively.

Possibly first discovered by Pythagoras, then systematically studied by the ancient mathematician Euclid and intuitively explained in Plato's *Timaeus* (Lučić 2009, 205), these polyhedra have common properties that set them apart from the wider polyhedral family. Their faces are regular congruent polygon surfaces, all vertices and faces are convex and lateral, their planes are convex polygon, and all vertices are convex. Different combinations of golden sections can be found in their cross-sections, considered an ideal of harmony in ancient Greece. The dodecahedron (with twelve faces) is closest in shape to a sphere and named as the divine image by Plato: "There is another, fifth form: the god used it for arranging the starry constellations in the skies." In other words, it corresponds to the firmament.

While the dodecahedron corresponds to some degree to the ancient Chinese idea of *tian* 天 (sky, heaven, celestial realm), it represents more a specific ancient attempt to deductively prove the existence of a perfect form or an Absolute (idea) as the source of all existence. The dodecahedron has twelve faces representing the zodiac and cosmic time.

Also, while in nature the tetrahedron, hexahedron, and octahedron occur during the crystallization of minerals, the creation of a dodecahedron and an icosahedron (which is a dual form of a dodecahedron), require a human hand or a transcendent Creator. That is why the dodecahedron has a special place in Plato's system: it is a moving image of eternity, an image of time itself, the first emanating principle of the creator.

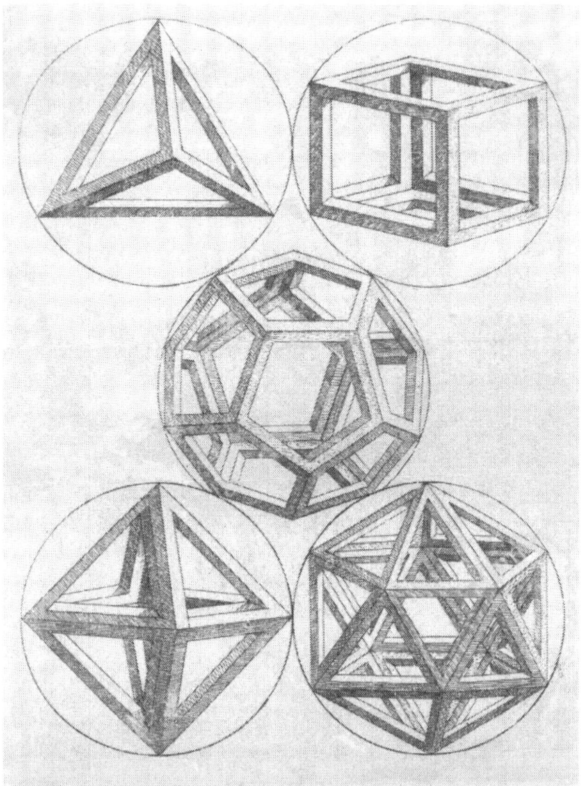

Close to the circle as a perfect form, the relationship between the two is complex and hides the secret of the squaring of the circle and irrational and incommensurable numbers. To find the mathematical solutions for their relationship means to find how the creator generates the world. First Plato then Euclid tried to explicitly connect mathematics and ontology. They pointed out, presenting a whole system of evidence, that the basis of the existence of beings is non-intuitive and mental.

In such a perspective, time clearly stands out as immeasurable emanation—part of the world of illusions explained in the myth of the cave. Defining time as a moving image of eternity, whose essence is expressed in the generation of life forms from basic and regular geometric shapes,

Plato tried to synthesize atomistic and Eleatic learning as the dominant and mutually opposite teachings of his time.

The Body in Time

Most of Plato's *Timaeus* focuses on explaining the creation and functioning of the human body, which also arises from a perfect sphere. Created in a similar way as the universe through the demiurge, its foundation is spherical, with the head as its central and most important part. Its extension into limbs and torso is due to the fact that, unlike the perfect sphere, it is not in a position to move evenly without resistance. Rather, it is influenced externally and thus gets sick, declines, ages, and dies—moving along a timeline that is both linear and downward oriented.

The body is further explained through an analogy with the net, which traps the soul in a deceptive way: the body is like a fish-coop that prey can enter but not escape, again emphasizing the one-way, entropy-bound nature of Western temporal thinking. Plato also calls it the dungeon of the soul. Since it works like a mechanism operated by the demiurge, it has self-regulating properties and, once triggered, gains a certain level of independence from its creator, moving steadily and inexorably in one direction only (Wood 1968).

Plato distinguishes between the mortal and immortal parts of the person. When the body dissolves at death, the soul abandons it, leaving for a higher, purer realm, which is beyond the flow of earthly time, a timeless state of the eternal One. He accordingly portrays life allegorically as a large cave—dark and determined by entropy, where the play of light and shadow is deeply illusory. Education and learning, then, cause people to turn to their timeless and spaceless essence, leading ultimately to an exit from the cave onto a plateau where truth manifests itself in eternal splendor. Pure light, utterly revealed, and free from shadow, it shows ideas themselves—beyond all vicissitudes of temporal existence—in a pure self-representational glow (Heidegger 1984). The transformative path of the soul, in other words, leads from the body and the timed to the disembodied, immortal, and timeless (Saunders 1973, 232-44).

In contrast, Daoists work thorugh self-cultivation toward a refined state, moving from grosser to subtler levels on one single energetic scale. Rather than freeing the soul from its imprisonment in the body toward a timeless eternity, they activate the body as a key player and supporter of spirit, reversing entropy and moving backward in the flow of time. Much of Daoist practice consists of using, controlling, conserving, transmuting, and circulating the body's natural energy flow in order to replace wear and tear, exhaustion and depletion with refined vital energy, recovering youthfulness, vigor, and increased vitality. In other words, it leads from disintegration to reintegration, from degeneration to regeneration. Daoists reverse rather than escape the timed existence of this world.

The body here is a place of microcosmic creation, representing the entire natural world on a smaller scale and similarly populated by deities, sacred animals, and palaces. Plato, according to the *Timaeus*, on the other hand, sees it as a geometric mechanism, a "box" crafted and produced by the demiurge. For him, the body is a cave full of illusions that can be overcome by dialectic and logical thinking, whereas in Daoism it contains not one but a multitude of caves and caverns, wells, streams, hollows, palaces, and all sorts of energy passages, through which the torrents flow and the winds bluster. It is filled with bubbling springs, hidden passages, wide roads, shining stars, and splendid paradises inhabited by immortals and expresses the Daoist ideal of attaining full realization through the refinement and transfiguration of the body in the ultimate unity of essence, energy, and spirit.

The Daoist approach to the body in time is works with the belief that flow can move in either direction, toward increased entropy and decline or toward rejuvenation and higher purity. The three levels or qualities of human existence—essence, energy, and spirit, jointly known as the "three treasures"—naturally transmute from the subtler to the grosser in the course of life and can accordingly also be transfigured in reverse order by refining their circulation and recovering more original, primordial levels. Just as in the magic square, the order of the three treasures is not linear, but all work together and circulate around each other in a variety of ways, allowing movement in all directions. Their main symbol is water, a key factor in transformative change, generally considered an image of supreme power, beauty, enlightenment, and even immortality, since it flows constantly, opposes nothing, and wins over everything. Like time itself, it purifies all and has no lasting form—making it also a symbol of freedom.

Conclusion

The comparison of Daoist and ancient philosophies, mathematics, and medicine points to some worldwide characteristics typical for the period from the 6th-3rd centuries BCE, such as merging myth with philosophy, magic with religion, reality with scientific interpretation. There is also a common interest in establishing order on the basis of the principles of cosmological harmony. The cultural differences between Daoism and Greek philosophy are most obvious in their reading of the concept of change, crucial for understanding both, Daoism and the perfect form that came to define ancient philosophy with Parmenides.

As a result, explanations of the principles of how cosmos and reality function in ancient China are inductive, while in ancient Greece they rest on the principle of deduction and logical proof. Ancient Chinese magic squares in comparison to the perfect sphere in ancient Greece reveal these differences poingnantly, as does the analysis of basic mathematical differences in their interpretation—all to help explain the variant trajectories in the development of Chinese and Western civilization.

Bibliography

Allan, Sarah. 1991. *The Shape of the Turtle: Myth, Art, and Cosmos in Early China*. Albany: State University of New York Press.

Boodberg, Peter. 1957. "Philological Notes on the Chapter One of the Lao Tsu." *Harvard Journal of Asiatic Studies* 20.3-4:598-617.

Božić, Milan. 2010. *Pregled istorije i filozofije matematike*. Beograd: Zavod za udžbenike.

Burik, Steven Victor. 2006. "The End of Comparative Philosophy and the Task of Comparative Thinking: The Language of Comparative philosophy, Seen through a Comparison of Martin Heidegger, Jacques Derrida, and Classical Daoism." Ph. D. Diss., National University of Singapore, Singapore.

Clegg, Jerry S. 1976. "Plato's Vision of Chaos." *The Classical Quarterly* 26.1:52-61.

Gill, Mary Louise. 1987. "Matter and Flux in Plato's 'Timaeus.'" *Phronesis* 32.1:34-53.

Graham, A. C. 1986. *Yin-Yang and the Nature of Correlative Thinking*. Singapore: Institute for East Asian Philosophies.

Hackforth R. 1959. "Plato's Cosmogony." *The Classical Quarterly* 9.1:17-22.

Heidegger, Martin. 1984. "Platonov nauk o istini." *Luča, časopis za filozofiju i društveni život* 1.2:5-34.

Kline, Morris. 1962. *Mathematics: A Cultural Approach*. Reading, Mass.: Addison-Wesley Publishing.

Kohn, Livia, ed. 2006. *Daoist Body Cultivation: Traditional Models and Contemporary Practices*. Cambridge, Mass.: Three Pines Press.

_____. 2008. *Introducing Daoism*. London: Routledge.

_____. 2021. *Taming Time: Daoist Ways of Working with Multiple Temporalities*. St. Petersburg, Fla.: Three Pines Press.

Kung, Joan. 1985. "Tetrahedra, Motion, and Virtue." *Noûs* 19.1:17-27.

Legge James. 1891. *The Writings of Chuang Tzu*. Chinese Text Project. www. ctext.org.zhuangzi.

Louis, Francois. 2003. "The Genesis of an Icon: The Taiji Diagram's Early History." *Harvard Journal of Asiatic Studies* 63:145-96.

Mair, Victor H. 1998. *Wandering on the Way: Early Taoist Tales and Parables of Chuang Tzu*. Honolulu: University of Hawaii Press.

Martzloff, Jean-Paul. 2016. *Astronomy and Calendars: The Other Chinese Mathematics 104 BC-AD 1644*. New York: Springer.

Miller, Harold W. 1957. "The Flux of the Body in Plato's *Timaeus*." *Transactions and Proceedings of the American Philological Association* 88:103-13.

Mohr, Richard D. 1978. "The Formation of the Cosmos in the 'Statesman' Myth." *Phoenix* 32.3:250-52.

Needham, Joseph, et al. 1954. *Science and Civilisation in China*, vol. I: *Introductory Orientation*. Cambridge: Cambridge University Press.

Niče Fridrih. 1984. *Rođenje tragedije*. Beograd: Filip Višnjić.

O'Neill, Timothy Michael. 2006. *Ideography and Chinese Language Theory*. Boston: De Gruyter.

Plato. 1966. *Država*. Beograd: Kultura.

_____. 1995. *Timaj*. Vrnjačka banja: Eidos.

Pregadio, Fabrizio. 2004. "The Notion of 'Form' and the Ways of Liberation in Daoism." *Cahiers d'Extrême-Asie* 14:95-130.

Pušić, Radosav. 2001. *Kosmička šara: O starokineskoj filozofiji*. Beograd: Plato.

Saunders Trevor J. 1973. "Penology and Eschatology in Plato's *Timaeus* and Laws." *The Classical Quarterly* 23.2:232-44.

Schuyler, Cammann. 1960. "The Evolution of Magic Squares in China." *Journal of the American Oriental Society* 80.2:116-24.

Schwartz, I. Benjamin. 1985. *The World of Thought in Ancient China*. Cambridge, Mass.: The Belknap Press of Harward University Press.

Sekulić, Nada M. 2016. "Janusovo lice mizoginije; imaginarne anatomije i političko-simbolička značenja žesnkog tela u srednjem veku i renesansi." *Antropologija* 16.1:53-71.

_____. 2015. "Istorija i estetika zmaja - o estetizaciji nasilja i problemu orijentalizma u Hegelovoj estetici." In*Aktuelnost i buducnost estetike*, edited by Zbornik radova Estetickog drustva Srbije, 297-31. Beograd: Esteticko drustvo Srbije.

Skemp J. B.1947. "Plants in Plato's *Timaeus*." *The Classical Quarterly* 41.1-2:53-60.

Strickland, Lloyd. 2006. *Leibniz Reinterpreted*. London: Continuum International Publishing Group.

Swetz, Frank. 1979. "The Evolution of Mathematics in Ancient China." *Mathematics Magasine* 52.1:10-19.

Temple, Robert. 1986. *The Genius of China: 3000 Years of Science, Discovery, and Invention*. New York: Simon & Schuster.

Verellen, Franciscus. 2006. "The Dynamic Design: Ritual and Contemplative Graphics in Daoist Scriptures." In *Daoism in History: Essays in Honour of Liu Ts'un-yan*, edited by Benjamin Penny, 159-86. London: Routledge.

Wang, Robin R. 2005. "Zhou Dunyi's Diagram of the Supreme Ultimate Explained (*Taijitu shuo*): A Construction of the Confucian Metaphysics." *Journal of the History of Ideas* 66.3:307-23.

Wang, Aihe. 2006. *Cosmology and Political Culture in Early China*. Cambridge: Cambridge University Press.

Watson, Burton. 1964. *Chuang Tzu: Basic Writings*. New York: Columbia University Press.

Wolfsdorf, David. 2014. "'Timaeus' Explanation of Sense-Perceptual Pleasure." *Journal of Hellenic Studies* 134:120-35.

Wong, Eva. 1992. *Cultivating Stillness*. Boston: Shambhala.

Wood, Richard James. 1968. "The Demiurge and His Model." *Classical Journal* 63.6:255-58.

Xu, Keqian. 2010. "Chinese 'Dao' and Western 'Truth': A Comparative and Dynamic Perspective." *Asian Social Science* 6.12:42-49.

Zoran Lučić. 2009. *Ogledi iz antičke geometrije*. Beograd: Službeni glasnik.

Resurrecting Daoist Temporalities in the Taiwanese New Wave

WUJUN KE

Contemporary Chinese directors such as Jia Zhangke 賈樟柯, Wang Bing 王兵, and Liu Jiayin 刘伽茵, as well as Taiwanese directors Hou Hsiao-hsien 侯孝賢 and Tsai Ming-Liang 蔡明亮, are often grouped together as contemporary Asian exemplars of "slow cinema," an international film style characteristic of arthouse and festival films. While slow cinema is frequently discussed in relation to critical theories of duration (such as Deleuze's time-image and Bergson's durational time), I aim to de-secularize discussions on cinematic temporality and use a Daoist lens to excavate the significance of the Daoist references they employ. By centering spiritual traditions within films that dilate time, I follow Walter Benjamin's call to complicate notions of secular, homogenous time with messianic time. In addition, I situate the wider significance of my argument in discussions about the "postsecular" in support of the idea that religion does not disappear with the onset of secular modernity, but instead remains an influential part of public life (Habermas 2008).

It is crucial to trace how spiritual traditions intervene in contemporary social and economic temporalities in order to define alternatives to the temporality of secular modernity. Secular modernity, since the advent of nation-states, has been dominated by time that is secular, homogeneous, chronological, and measured by the clock. Produced alongside capitalism, it is used to regiment life for the increasingly brutal extraction of labor. In concrete terms, modern time is divided into measurable quantities: days, weeks, months, and years before and after the year zero. This quantitative, chronological system grew into an official means of marking time, spreading globally in tandem with European colonial ventures and Hegelian notions of teleological history.

By contrast, Daoist notions of time embrace relativity and multiplicity, positing that the world exists in continual transformation. Rather than regimenting time into compartmentalized units of clock-time or a historicist succession of past, present, and future, Daoists subscribe to a notion of time as unfolding in the interconnected context of the natural world, going far beyond the limited confines of the human world. The temporality of Daoism advocates for a return to multiplicity and contingency prior to the Way's distortion by the social conventions of the human world.

In this essay, I situate a post-secular approach in the context of Taiwan's post-martial law period, a time of rapid neoliberal transformation documented by Taiwanese New Wave filmmakers. I discuss Edward Yang's *Terrorizers* (*Kongbu Fenzi* 恐怖份子, 1986) as well as his landmark work *Yi Y* 一一 (A One and a Two, 2000) plus Tsai Ming-Liang's *Stray Dogs (Jiaoyou* 郊遊, 2013), in order to meditate on how they invoke Daoist attitudes towards time, process, and transformation. Each exemplifies the New Wave style in its observational realism and focus on ordinary life. I select these two films as exemplary of two possible directions that Daoist images of time might take. While *Yi Yi* is attuned to the cyclical, nourishing, self-renewing rhythms of everyday life, *Stray Dogs* emphasizes the impartiality of nature and inevitability of loss. Both films resurrect Daoist temporalities as interventions on the homogenous, empty time of late capitalist modernity.

Situating the Taiwanese New Wave

Laden with a multiplicity of sometimes contradictory signifiers—as a repository of traditional Chinese culture, an outpost for global techno-capitalism, a former Japanese colony, home to at least sixteen aboriginal groups, and an island under increasing pressure from China to relinquish its political sovereignty, Taiwan can be thought of a site of crisis, invisibility, and liminality. Located off the coast of China between the East and South China seas, it has been subjected to multiple waves of colonization, including by the Spanish, the Dutch, Ming loyalists, and the Japanese Empire. Because of these intersecting political erasures, Huang argues, Taiwan embodies a structure of "negative presence," referring to the "historical condition of subjection to multiple systems of inclusive exclusions" (2020, 189).

Taiwan's modern governance began when the Kuomintang (KMT) retreated from mainland China in 1947 to take temporary refuge in Taiwan before, as they planned, returning to reconquer the mainland. The KMT was extremely repressive, jailing and killing thousands of suspected communists in what became known as the February 28 incident, launching the "White Terror" and four decades of martial law. In 1979, President Carter recognized the People's Republic of China and cut off relations with Taipei, setting a precedent for other global powers to do the same. In the 1980s, massive protests led to the lifting of martial law and the beginnings of democratization. During the 1970s and 1980s, Taiwan's adoption of an export-driven economy, moreover, led to an intensive period of industrialization and the growth of an urban middle class.

After martial law was lifted, Taiwanese New Wave filmmakers took advantage of the increased political openness and committed themselves to a realist mode that portrays the impact of whirlwind political, economic, and social developments on Taiwan's psychic life. They also benefited from the cosmopolitan offerings of a KMT regime that sought global legitimacy, the low-budget filmmaking of the Italian neo-realists, as well as rebellious rock music from the United States. Because political ideals were not allowed open expression in the authoritarian social environment; however, Edward Yang has described the psychic life of his generation as characterized by "inner rage and outward conformity" (2001, 130). This sense of tension between inner turmoil and external placidity plagues the protagonists in one of his landmark films, *Terrorizers*.

Popularized by literary theorist Frederic Jameson as a touchstone of postmodern cinema, *Terrorizers* remains one of the most recognizable films of the Taiwanese New Wave. The film is organized around at least two unrelated storylines that intersect in unpredictable ways: a wealthy young photographer's voyeuristic obsessions leads him to the scene of a crime, where he takes pictures of a Eurasian girl ("White Chick") in the middle of a hasty escape. Meanwhile, Li Li-Chun, a medical professional in line for a promotion, navigates a strained relationship with his wife, a novelist battling writer's block and domestic malaise. Each of the characters is listless, bored with his or her life, and starved for novelty and excitement. Their storylines intersect when the Eurasian girl makes a prank phone call, a minor event with devastating consequences.

Frederic Jameson's reading of *Terrorizers* defined the film as a Third World national allegory situated within the postmodern world system. He argues that *Terrorizers* appropriates a modernist storytelling technique, Synchronic Monadic Simultaneity, to link storylines that have no apparent relationship to one another. Jameson interprets this tendency as evidence for the globalization of postmodern subjectivities, which are no longer tied to moral categories of good and evil, but instead produced by capitalist processes of commodification and reification. Jameson notes that, unlike earlier films in the Taiwanese New Wave, there is an absence of worries about the nature of Taiwanese identity in *Terrorizers*. Rather, he sees the main character, Li Li-Chung, as a "quintessential loser" and an allegory for a post-third world country that can never really join the first world, an embodiment of "fantasies about the limits to Taiwanese development in a world system" (1995, 146).

Scholars have taken Jameson's reading of *Terrorizers* to task, arguing that such a reading fails to account for the specificity of Taiwan's history as well as New Wave directors' desires to write stories grounded in local perspectives. Liu points out that, rather than serving as allegories for Taiwan, many of Edward Yang's male characters are out of step with the country's rapid pace of modernization and economic development (2017, 116). Wu Yuyu notes that Jameson's focus on globalization in *Terrorizers* minimizes the longer historical context in which Taiwanese New Wave directors in the 1980s were dedicated to rewriting Taiwan's history from an indigenous perspective (2020, 56).

After the 1990s, indigenization was entangled with globalization and could no longer be separated into distinct historical periods. Instead, he argues, the film is situated in a complex, overlapping context with loneliness as a consistent psychological pattern for Taipei throughout different historical periods. Instead of attributing the characters' psychic alienation to the spread of postmodernism, Wu cites Wu Zhuoliu's 吳濁流 *Orphan of Asia* (*Yaxiya de gu'er* 亞細亞的孤兒) as a reference point for a psychic loneliness that stretches back at least as far as the Japanese colonial era.

I also seek to complicate linear conceptions of history by suggesting that traditional modes of thought, such as Daoism, continue to exist well into modernity and post-modernity. Next, I will discuss Edward Yang's final film, *Yi Yi* (2000), as illustrative of my post-secular argument.

Yi Yi

Unlike the self-centered protagonists of *Terrorizers*, the protagonists of Edward Yang's *Yi Yi* are deeply sympathetic. By indulging in long takes and leisurely pacing, the film gives its characters and viewers the time to think and reflect, offering an experience of lived time that is deeply mundane yet magical in its simplicity. Translated into English as "A One and Two," the title conjures the image of a conductor waving his baton before launching into a performance. The film is an orchestral masterpiece that weaves together many melodies of the Jian family's life in 21st-century Taipei.

It follows three family members: the soft-spoken patriarch N. J., the growing teenager Ting Ting, and the budding artist Yang Yang. They live in a small apartment with their mother and maternal grandmother, who has an accident and falls into a coma right in the beginning of the film. The family's apartment—cozy, lived-in—becomes a gathering ground for family members and the holding space for the comatose grandmother. It is inside, outside, and from the third-floor window of the building's balcony that the characters' lives play out.

The film portrays Taipei at the turn of the millennium—in the midst of socioeconomic change, struggling to compete in the global economy while negotiating with traditional values informed by Confucian, Daoist, and Buddhist thought. This negotiation is personified in N. J.'s struggles to be honest with his clients at work under the increasing sway of the profit motive. His struggle to preserve his integrity in dealings with a Japanese client, Mr. Oto, is threatened by his coworkers' desire to maximize profit by using cheaper software produced by a copycat company in Taiwan. This filmic sub-plot writes an alternative story of what has been called Taiwan's "economic miracle"—a boom period resulting from its transition in the mid-1980s from state-led industrialization to an export-driven, globalized economy. During this time of rapid change, this slice-of-life film momentarily suspends larger historical trends and zooms into the joys and frustrations of one Taiwanese family.

While the film traces familiar New Wave motifs of urban alienation and loneliness, it also seeks answers to existential questions—the meaning of life and death, the passage of time, and how spiritual traditions morph and adapt to a materialistic society. The film makes abundant use of the long take to highlight the understated moments of everyday life. Because the camera lingers on the urban landscape as punctuation marks or moments of pause, it makes the viewer feel that bland, unremarkable moments in life are equally as important as the twists and turns of an action film. While the film is not short of drama—including a murder, a suicide attempt, a hasty marriage, infidelity, and a funeral—events are subordinated to quieter moments meant to illustrate the durational flow of time. Scenes move gracefully and music sometimes spills over the frames. Life is constantly happening on the margins of the narrative and in still moments between plot motion, inviting the viewer to travel into a meditative plane by way of the mundane.

It is difficult to put into words the reasons why this film is so moving. As A.O. Scott of *The New York Times* reflects, "I struggled to identify the overpowering feeling that was making me tear up. Was it grief? Joy? Mirth? Yes, I decided, it was all of these. But mostly, it was gratitude" (2000). *Yi Yi*'s beauty seems to lie in its ability to capture something ineffable, making the film a challenge for critics and scholars to decipher.

According to Gilles Deleuze, these strong yet unnamable feelings can be described as affect, a kind of movement away from preconceived emotions—a kind of emotional deterritorialization—and thus difficult to describe in language (1986). Just as film uses visual language, narrative, and editing to articulate what cannot be fully captured in words, I posit that Daoist philosophy is uniquely positioned to articulate the ineffable quality of this film, as it is characteristically skeptical of language. It also embraces a more intuitive form of knowledge tied to a temporality that emphasizes the process of unfolding.

Time is suspended in many ways in the film, but most significantly in the unanchored temporality of the comatose grandmother. In the beginning, Ting Ting, the teenage daughter, forgets to take out the trash. This mistake leads to the grandmother taking out the trash, tripping on the stairs, and falling into a coma for the rest of the film. A doctor advises the family to speak to the grandmother every day to help her recover, and the family members take turns speaking in front of her bed without receiving a response. Ting Ting feels guilty over her forgetfulness having led to her grandmother's coma and begins to suffer from insomnia.

Near the end of the film, Ting Ting finally dozes off and enters into a lucid dream in which her grandma gives her an origami butterfly. This scene references Zhuangzi's famous butterfly dream, which imparts the wisdom that dreaming and waking, life and death, are merely transformations from one state to another:

> Once Zhuang Zhou dreamt he was a butterfly, a butterfly flitting and fluttering around, happy with himself and doing as he pleased. He didn't know he was Zhuang Zhou. Suddenly he woke up and there he was, solid and unmistakable Zhuang Zhou. But he didn't know if he was Zhuang Zhou who had dreamt he was a butterfly, or a butterfly dreaming he was Zhuang Zhou. Between Zhuang Zhou and a butterfly there must be some distinction! This is called the transformation of things. (Watson 2013, 99)

One is not necessary more real or more significant than the other, and one should strive to approximate the inclusive impartiality of the cosmos and embrace transformation as it occurs. The story illustrates the Daoist principle of the "equality of things" (*qiwu* 齊物)—a sense of equilibrium and willingness to let go of attachments to temporary states of being. The reference to the *Zhuangzi* serves to open up a more cosmological space that allows Ting Ting to accept the passing of her grandmother as simply part of the natural and inevitable transformations of life and death, just as waking transforms to dreaming and vice versa. She mysteriously finds an origami butterfly in her palm upon waking, a reminder of her dream of forgiveness and release from her guilt.

Messianic Time

The Zhuangzian butterfly in *Yi Yi* stems Ting Ting's feelings of guilt by warping linear temporality and opening up a space for rest and contemplation. It also creates a thread of messianic connection between the past and present as well as between the spiritual and secular realms.

In Ting Ting's dream as in her waking life, the butterfly is a messianic figure that redeems her from her mistakes and her guilt over those mistakes. It interrupts ordinary life and opens up a temporality of fulfilled hope, what we might borrow from Walter Benjamin and call messianic time. Writing at the end of World War II, Walter Benjamin celebrates the messianic in "Theses on the Philosophy of History."

Rejecting the Hegelian idea that history obeys inexorable laws allowing for the attainment of ever-higher collective consciousness, Benjamin highlights that the past is contingent and fleeting. History's continuity is that of the oppressors, while history of the oppressed is fragmented and discontinuous. For Benjamin, it is the historian's task to interrupt progressive time by blasting moments out of the continuum of history and forcing time to a critical stand-still. This allows us to recognize moments of immediacy, what he calls "now-time" (*Jetztzeit*), which fulfills and momentarily suspends the progression of history as an ongoing catastrophe.

"Now-time" serves as an alternative to linear, historicist temporality. It is often produced by moments of interruption, unexpectedness, and wayward-ness, and in *Yi Yi*, this waywardness is personified by the youngest character, Yang Yang. Yang Yang is a misfit, constantly teased by the older girls around him and getting into trouble at school. When his father gifts him a camera, he begins to experiment with photography and through this practice, develops a deeper and more philosophical understanding of life.

To the bafflement of those around him, he likes to take photos of the backs of people's heads. He explains to his father, "You can see what I can't see, and I can see what you can't see, so how do I know what you can see? . . . Aren't we only able to see half of something?" The photos he takes freeze a moment in time, as if literalizing Benjamin's imperative to bring history to a critical stand-still. Despite his young age, Yang Yang also seems to serve as a messianic figure in the film, seizing through his camera the essence of "a memory that flashes up at a moment of danger" (Benjamin 1969, 255). While his practice is grounded in aesthetics rather than politics, he aspires to show people that the social totality is composed of a combination of partial perspectives. Using an ordinary example of the impossibility of seeing the back of our heads, Yang Yang raises the question of how we can say for sure that any of our singular perspective or values are true. This movie hints that only through skepticism and openness to unconventional framings can we open our eyes to this expansive and multidimensional world, echoing the relativism and pluralism of Daoist thought.

People make choices and must accept the consequences of their actions, the film tells us. Those who do not accept what life throws at them look a bit lost and silly, and no one who clings to worldly success or stability is happy. At the end of the film, each family member arrives at a sense of spiritual acceptance. While meeting up with his first love in Japan on a business trip, the father N. J. reminisces over the regrets and longings of his youth in his middle-age. He does not, however, stray into the escapist cliché of a man undergoing a mid-life crisis.

Despite being offered a fresh start with an ex-lover, someone he claims is the only woman he has ever loved, N. J. rejects her advances and returns home to his wife and family. He confesses to his wife about meeting his first love on the business trip, reflecting: "It's just that I suddenly realized, if I were to start over . . . there's really no need." His acceptance of what he has experienced and lost, with no desire to start over, is an example of following the Way. He draws a striking contrast to the characters in *Terrorizers* who would love nothing more than to start over again and again, each time without finding what they are looking for.

Modes of Duration

Malaysian-born Taiwanese filmmaker Tsai Ming-Liang's *Stray Dogs* (2013) is most obviously notable for its excruciating, nearly 12-minute take near the end of the film. It is considered one of the quintessential examples of slow cinema, a group often labeled as "festival films" because they are generally considered too austere even for arthouse distribution and exhibition. *Stray Dogs* illustrates key characteristics of slow cinema, including the frequent use of the long take, undramatic or non-narrative structure, realist or hyperrealist representation, and a marked stillness of the camera as well as its content (Flanagan 2012; Lim 2014).

Flanagan in particular identifies the historical origins, from which slowness emerges as the trauma of World War II, a feat that accords with Gilles Deleuze's distinction between the movement-image and the time-image in *Cinema 2* (2012). Responding to the emergence of fast-paced commercial cinema and mass advertising, slowness became a way for art filmmakers to depart from the dominant strategies of Hollywood contextualization and to define themselves against classical narrative models.

Beyond merely contravening dominant film temporalities, however, slow cinema can also be a means to reflect on the accelerationism of global capitalism and critique widespread mental habits of instantaneous gratification fostered by contemporary media technologies.

As implied by its name, one of the main characteristics of slow cinema is the "hyperbolic application of the long take" (Luca 2011, 12). This formal property produces a mode of realism that requires observational study rather than immediacy of action and reaction. Because deferrals of understanding are built into the structure of the film for both the onscreen characters and the audience, the long take could be considered a specific manifestation of what Gilles Deleuze calls the "irrational cut."

What characterizes a long take is not merely the time elapsed between cuts, but the deferral of rational continuities. In the archive of films that Deleuze defines as "images of time," these films reject a predictable editing structure in favor unusually long takes and unpredictable linkages from one shot to the next. Rejecting the narrative logic of classical cinema's privileging of movement over time, they seek to extract a direct image of time. According to Deleuze, "It is no longer time that depends on movement; it is aberrant movement that depends on time" (1986, 41).

Deleuze's account of the time-image draws from Henri Bergson's understanding of time as qualitative experience. In contrast to standardized clock-time, which is homogeneous for every place on earth, Bergson's durational time theorizes heterogeneity as the ground of lived experience. Change and continuity coexist in a vibrational universe composed of force relations. Embracing a philosophy of intuition, Bergson affirms that temporal multiplicity is lived, not thought. Bergsonian duration is an open, relational whole, and never given or determined ahead of time. Rather than supporting historical closure through projections about the future, it affirms the fact that nothing is exactly what we thought it was, and nothing is what we think it will be.

Contemporary scholars have been reawakened to Bergson's ideas through Deleuze's primer *Bergsonism* (1990, orig. 1966) and his two volumes on cinema. In *Bergsonism*, he calls for a return to Bergson—not only to return to his ideas, but also to extend his project in relation to the contemporary transformations of science and society. Deleuze turned to Bergson to escape the Hegelianism that was prevalent in the French academy. In contrast to the abstraction of the Hegelian dialectic, Bergson's philosophy of lived experience affirms internal difference, novelty, indetermination, and unforseeability.

In *Time and Free Will* (2018, orig. 1889), Bergson theorizes real duration on the phenomenological level—defined as the inner experience of time in its immediacy, independent of space and inaccessible to cognitive consciousness. The ground of experience is qualitative richness of sensations or qualitative multiplicity, in which states of mind "overlap, merge with one another, and add together dynamically, forming a qualitative, or 'confused,' multiplicity" (Guerlac 2006, 96). Thinking interiority and subjective experience means that in the process of becoming, we never feel the same thing twice.

Duration goes beyond the phenomenological, as Bergson also establishes a permeability between the realm of inner experience and the nature of external reality. In *Introduction à la métaphysique* (2011, orig. 1903), he theorizes reality not as a series of discrete states but a process of constant change and ever-fluid movement. Reality can only be seized through intuition, or "entering into" the object, rather than intellectual analysis, which can never go beyond mental shortcuts of homogenization and quantification.

In *Matter and Memory* (1991, orig. 1896), he further clarifies that the mental utilitarianism of human consciousness performs processes of abstraction, solidification, and division on the "moving continuity of the real in order to obtain a fulcrum for our action" (1991, 280-82). While useful for calculative tasks, these mental shortcuts—homogenization, quantification, abstraction—spatialize time into a falsely linear chronology. Similarly, language can never quite capture the kaleidoscopic, moving continuity of duration, in which "everything changes and yet remains" (1991, 260). For Bergson, there is no dissociation between permanent bodies and homogenous movements in space, for they are part of the same kaleidoscopic whole. Rather than thinking about individual bodies with determinate boundaries that move in empty space, he speaks of change as forces, "modifications, perturbations, changes in tension or energy"—metaphors drawn from fluid dynamics that Gilles Deleuze and XXX Foucault would later take up in their philosophical thought (1991, 226).

While sensory-motor perception reduces reality to representations that serve our interests and needs, attention or attentive recognition is

> a turning back of the mind which gives up pursuing any useful end of the present perception: there will be first an inhibition of movement, a pause. But rapidly, other, more subtle movements will graft themselves onto this attitude . . . whose role it is to go over the contours of the perceived object. With these movements the positive work of attention begins, and not just the negative work. It is continued by memories. (1991, 110)

Attention is anti-utilitarian, though it also appeals to memory. An increase of attention means a widening scope of memory images, allowing the image at hand to become embedded in a constellation of past images. It increases our ability to contextualize the perceived image within deeper layers of memory images, allowing us to resolve objects with greater detail and specificity than utilitarian, action-oriented perception. Illustrating his ideas using the diagram of a cone of memory, whose point intersects with the plane of perception, the present moment (represented by the point of intersection) is the moment of useful action. The cone represents pure memory. The more we move toward the more expanded intervals of cone and the expanded plane of perception, the more distended our attention and less focused we are on action. *L'esprit* (the mind) is in constant motion across these levels or intervals, while the body anchors our attention to the preservation of life.

The experience of attentive recognition requires drifting away from the pragmatism of the sensory-motor schema, whose focus on the utilitarian present necessarily fashions the chaos of time into a sensible plot. Hence, slow cinema involves the technical application of extreme long-takes and static framing to construct an experience of attentive recognition, an experience in which the virtual—pure memory and pure perception—is emphasized over the actual. Only when narrative framings are abandoned does durational present emerge, and this visualization of duration makes viewers aware of the expansiveness of a reality without any tangible reference points. Undoing established epistemologies as well as ways of seeing, it forces viewers to become familiar with the freedom that exists prior to language and the vulnerability that such freedom entails.

The static long take forces us to endure through our impatience, to experience the discomfort of real novelty, to not know in advance what is going to happen next on screen, to become completely open to any tiny shifts of movement, brightness, or mood in the diegesis. We are forced to release our conventional ways of thinking and living without immediately mapping novelty onto pre-ordained coordinates, rationalizations, or explanations. Laura Kissel describes the long take as "a process of discovery, enabled by the duration of the frame." She compares the long take to an investigation, and experiment, "a moment to linger on a question rather than pursue a particular answer" (2008, 351). Because slow cinema largely dispenses with narrative organization, we cannot fall back on the ordinary formulas of making order out of chaos. If we are bored, it is because we are overwhelmed and unused to being in a completely open, expansive field with no guide posts for where to look next.

Stray Dogs

I take the notion of the long take as a capacious one—beyond describing only the length of a single shot, it can also encompass a filmmaker's obsession with time and its representation on the screen. For instance, Tsai Ming-Liang returns to the same actors repeatedly in his films as if documenting time through his feature films.

Starting in 2012, Tsai developed a slow walking performance with his lifelong collaborator, Lee Kang-sheng, into a series of short films set in different cities, including *No Form, Walker, Sleepwalk, Diamond Sutra,* and *Journey to the West.* This series of "walking meditations" features Lee Kang-sheng dressed in a monk's scarlet robe walking at a glacial pace in the bustling streets of Taipei, Hong Kong, Kuching (in East Malaysia), Marseille, and Tokyo. Scholars have made the connection between these walking meditations and Buddhist notions of temporality, suggesting that the protagonist is actualizing the pilgrimage of the character of the monk Xuanzang, who made the journey from China to India to bring back Buddhist sutras in the 7th century (Lim 2017).

Tsai Min-Liang identifies as a Buddhist and claims that his artistic work "possesses a spiritual component" (Vagenas 2013)· While scholars have pointed out the Buddhist references in his work, fewer have discussed the Daoist influences in his films. I build on the discussion of spirituality in his work by investigating the Daoist temporal frameworks in his self-declared last feature film, *Stray Dogs* (2013), which took home the Grand Jury Prize at the Venice Film Festival in 2013.

The film paints a psychological and sensorial portrait of Hsiao Kang, a single father attempting to raise his children on the outskirts of Taipei. We see the man and his two children eating amidst weeds and on public benches, washing up in public bathrooms, relieving themselves in tall grass and by abandoned buildings. The father has a job holding a sign advertising luxury real estate while he and his children live in a one-bedroom shack. After the father suffers a mental breakdown, a woman who regularly feeds stray dogs nearby takes the children into her care.

What happens next is uncertain: the sequence might belong to a memory, fantasy, or dream. We are inside of an apartment and it is the man's birthday, and his wife and children sing for him. Despite the baroque lighting and dirt-streaked concrete walls, this looks like a happy and intact family. They have a recliner, tables, chairs, and seem to live a comfortable middle-class existence. The signs of trouble are subtle, almost undetectable—perhaps they lie in his hidden flask of alcohol or the way that she flinches when she sees him. These subtle details gesture toward something sinister lurking under the surface; perhaps alcoholism or domestic violence.

In the last scene of *Stray Dogs*, the camera holds a static shot of a man standing behind a woman for twelve minutes in front of a landscape mural. Silence. Nothing happens for one minute, then two, then ten. Finally, twelve minutes pass and eventually the man drinks from a flask while a single tear rolls down the woman's cheek. He approaches and rests his head on her shoulder, while she stares ahead, seemingly unmoved. Eventually, she leaves. After a few minutes alone, he too exits the frame. This shot is the longest in a film composed almost entirely of static long takes and sparse dialogue.

Perhaps this scene signals the end of their marriage, but we are not told the two characters' personal history with one another, nor are we entirely certain about the time frame when this scene occurs. If conventional ideas about marriage and kids point to faith in a future, then the film skips forward to the point after durable intimacy unravels along with the clarity of chronological and causal linkages.

Disposability

In the film, three actresses play the part of a woman, but we are never certain if they play one and the same woman or several different women. She is psychologically inaccessible and always in the midst of leaving. Abandoned like the stray dogs that inhabit the wilderness alongside them, the man and his children live moment to moment, too precarious to make plans for the future. While the film may seem to take a humanistic view of the stray dogs and humans, evoking compassion for those society views as disposable, Tsai himself gestures to a different conception of disposability—one that is tied not to the capitalist economic system that generates people as disposable, but rather the cosmic impartiality of nature that sees all things, in a way, as disposable. In an interview with Charles Tesson, artistic director of the Critic's Week in Cannes, Tsai reflects,

> When shooting this film, I often thought of one expression from Laozi, "Heaven and earth do not act out of benevolence; they treat all things as sacrificial straw dogs" [*Daode jing* 5]. Those poor people and their children seem to have been abandoned by the world, but they still have to live. On the other hand, those who have power and influence seem to me to have forgotten about this world. They work incessantly on never-ending construction, but they do not know when destruction will arrive. (2015)

The line he cites from the *Daode jing* is usually interpreted to mean that heaven and earth follow non-human rules, i.e., they do not subscribe to the Confucian virtue of benevolence (*ren* 仁). They treat all things as "straw dogs" (*zougou* 芻狗), ceremonial objects used in ancient sacrificial rites. According to Ames and Hall, "These sacrificial objects are artifacts that are treated with great reverence during the sacrifice itself, and then after the ceremony, discarded to be trodden underfoot" (2003, 206). While this vision of disposability might strike modern readers as inhumane, cruel, or unkind, it also points to a vision of the Dao as being completely impartial in its treatment of all things. Because nature is indifferent to humans and human notions of morality, it treats us with the same degree of care or carelessness as it would non-human beings. Chapter 5 of *Daode jing* then goes on to discuss how the sage also treats all people with impartiality, without giving preference based on artificial codes of morality or other social values.

By invoking the ancient Daoist text, Tsai Ming-Liang avoids a critique of disposability based on human conceptions of social justice. As Erin Huang notes, the film's dispensing of narrative storytelling reminds us of "the unassimilability of precarious lives in a conventional humanizing framework" (2020, 194). Rather, Laozi advocates an encompassing vision that takes the position of existential humility, being willing to accept one's powerlessness and surrendering to change: "In the natural cycle, all things have their moment, and when that moment passes, they must pass with it. There is nothing in nature, high or low, that is revered in perpetuity." (Ames and Hall 2003, 149). The understanding is based on the premise that everything and everyone is disposable from the perspective of nature, even though humans create arbitrary hierarchies of disposability based on wealth, power, status, or prestige. Tsai gestures at how the poor are more vulnerable and closer to this fundamental truth of our inherent disposability, while the wealthy find ways to shield themselves from the lives of the poor as well as their own inherent fragility.

Using the impersonal lens of the *Daode jing* produces a reading of Tsai's long take not as a humanizing shot that seeks to bestow the dignity on oppressed people, but rather as a spiritualized image of time. The twelve-minute take lends itself to reflections on the span of the couple's relationship as well as that of the cosmos, inviting viewers into the Bergsonian experience of attentive recognition. The two characters gazing at the mural, mesmerized, are no longer looking at the art work itself but at the memories and fantasies it generates, taking in the demise of their relationship simultaneously with the beauty of the artwork and the idealized landscape it portrays. Absorbed by their reflections over the mountains and rivers depicted in the mural, the estranged couple stand upon a post-apocalyptic landscape of a building in ruins. As Tsai reflects:

> On one whole wall of a crumbling house the familiar landscape in Taiwan is portrayed in charcoal pencil. It is like standing in front of a mirror, looking at the farther shore in the mirror. It is both real and surreal. It is both close at hand and far away on the horizon. If everyone has an ideal world in their heart, a perfect farther shore, a place deep in their soul: isn't it right there?" (Tesson 2015)

Through close analysis of *Yi Yi* and *Stray Dogs* as films inspired by and speaking to Daoist worldviews, I contribute to the post-secular project of recovering religiosities embedded in a film tradition usually analyzed in secular terms: on the one hand, postmodernism and globalization, and on the other, indigenization and localization. Rather than subscribe to the binary notion of one or the other, I bring attention to the role of the Taiwanese New Wave directors Edward Yang and Tsai Ming-Liang in resurrecting Daoist thought in moments of narrative stillness. References to Daoism in these films serve as a tool for temporal dilation as well as ethical reflection on contemporary social problems, as well as a vantage point from which to critique dominant temporalities and frame lived time as much more fluid, entangled, and plural than we previously imagined. They share the insight that lived time is never entirely colonized by hegemonic systems, and instead, suffused with spiritual resilience and attuned to the cyclical rhythms of more expansive and cosmological ways of thinking.

Bibliography

Ames, Roger T., and David L. Hall. 2003. *Daode Jing: "Making this Life Significant" – A Philosophical Translation*. New York: Ballantine Book.

Benjamin, Walter. 1969. *Illuminations: Essays and Reflections*. Edited by Hannah Arendt. Translated by Harry Zohn. New York: Schocken Books.

Bergson, Henri. 1991 [1896]. *Matter and Memory*. Cambridge, Mass.: MIT Press.

_____. 2011 [1903]. *Introduction à la metaphysique*. Paris: Presses Universitaires Françaises.

_____ 2018 [1889]. *Time and Free Will*. Edited by Taylor Anderson. New York: Odin's Library Classics.

Deleuze, Gilles. 1990 [1966]. *Bergsonism*. Translated by Hugh Tomlinson and Barbara Habberjam. Princeton: Zone Books.

_____. 1986. *Cinema 2: The Time-Image*. Minneapolis: University of Minnesota Press.

Flanagan, Matthew. 2012. "'Slow Cinema': Temporality and Style in Contemporary Art and Experimental Film." www. ore.exeter.ac.uk/repository/handle/10036/4432.

Guerlac, Suzanne. 2006. *Thinking in Time: An Introduction to Henri Bergson*. Ithaca, NY: Cornell University Press.

Habermas, Jürgen. 2008. "Notes on Post-Secular Society." *New Perspectives Quarterly* 25.4: 17-29. www.doi.org/10.1111/j.1540-5842.2008.01017.x.

Huang, Erin Y. 2020. *Urban Horror: Neoliberal Post-Socialism and the Limits of Visibility*. Durham: Duke University Press.

Jameson, Fredric. 1995. *The Geopolitical Aesthetic: Cinema and Space in the World System*. Indiana University Press.

Kissel, Laura. 2008. "The Terrain of the Long Take." *Journal of Visual Culture* .3:349-61. www.doi.org/10.1177/1470412908096341

Lim, Song Hwee. 2014. *Tsai Ming-Liang and a Cinema of Slowness*. Honolulu: University of Hawaii Press.

Liu, Catherine. 2017. "Taiwan's Cold War Geopolitics in Edward Yang's The Terrorizers." In *Surveillance in Asian Cinema*, edited by XXX, PAGE NOS. London: Routledge.

Scott, A. O. 2000. "Of Taiwan's Bourgeoisie and Its Affecting Charms." *The New York Times*.www.nytimes.com/2000/10/04/movies/film-festival-review-of-taiwan-s-bourgeoisie-and-its-affecting-charms.html

Tesson, Charles. 2015. "An Interview with Tsai Ming Liang." *New Wave Films*.

Vagenas, Maria Giovanna. 2013. "Filmmaker Tsai Ming-Liang Says His Work Should Be Appreciated Slowly." *South China Morning Post*, August 27. www.scmp.com/lifestyle/arts-culture/article/1299497/filmmaker-tsai-ming-Liang-says-his-work-should-be-appreciated.

Watson, Burton, 2003. *Zhuangzi: Basic Writings*. New York: Columbia University Press.

Wu, Yuyu. 2020. "Remapping Taipei for Jameson? Rediscovering the Indigenization, Modernity, and Postmodernity of Taipei." *Critical Arts* 34.2:55-68. www.doi.org/10.1080/02560046.2020.1750446.

Yang, Edward. 2001. "Taiwan Stories." *New Left Review* 11:129-36.

Films

Edward Yang, *Terrorizers*. 1986.

Edward Yang, *Yi Yi*. 2000.

Tsai Ming Liang, *Stray Dogs*. 2013.

Ways of Time

Seeing Dao through Guénon, Teilhard, and Suzuki

PATRICK LAUDE

Daoism presents visions of time both cyclical and linear. According to its philosophy, the world manifests in cycles of flows and returns. The religious perspective, on the other hand, centers on a millenarian vision where Lord Lao leads the world toward a state of cosmic and social unity known as Great Peace. This raises first the important question of the relationship between spiritual transformation and cosmic change. At the same time, though, this vision presupposes a prior state of degradation resulting from the collapse of the spontaneous oneness with Dao.

Second, the more general question arises whether the religious view of time is descending or declining to culminate in final destruction or ascending toward a universal apogee. These two views can be characterized as entropic versus progressive.

Third, moreover, spiritual perception can also focus on transcending time and reaching a timeless eternity independent from the vicissitudes of personal life and cultural history. Many metaphysicians and mystics claim that this transcendent sense can be accessed in the present. The two first views are diachronic, while the third is both synchronic and trans-chronic.

The following pages explore some implications of this triad of aspects of time in philosophical Daoism, relating them to the works of the French thinker René Guénon (1886-1951), the evolutionist theologian Pierre Teilhard de Chardin (1881-1955), and the Japanese Zen philosopher D. T. Suzuki (1870-1966). It is significant that both Guénon and Suzuki approach Daoism as outsiders, while their respective intellectual and spiritual perspectives show strong affinities with this tradition.

To begin, Guénon contemplates Daoism from the point of view of a "perennialist" *Weltanschauung*, as one of the branches of the primordial Tradition. His most extensive metaphysical contributions root his work in Hinduism, as witnessed by his two classics, *Introduction to the Study of the Hindu Doctrines* (1921) and *Man and His Becoming according to the Vedānta*

(1925), while his own spiritual quest found its final abode in the Sufi path. However, several of his books deal extensively with Daoist concepts, most specifically *The Great Triad,* first published in French under the title *La Grande Triade* in 1946.

Guénon's interest in Daoism manifests in metaphysics, cosmology, and symbolism. He sees Confucianism and Daoism as two branches of the Chinese tradition harking back to Emperor Fuxi 伏羲 and the *Yijing* 易經 (Book of Changes). For him, Daoism is "Chinese esotericism," while Confucian teachings represent the exoteric side of the Chinese tradition. Precisely because of this difference, he sees Confucianism as liable to historical decline and disappearance, while Daoism remains out of touch from temporal demise. As he says,

> Confucianism, which represents only the exterior aspect of the tradition, might even disappear should social conditions happen to change to the point of requiring the establishment of an entirely new form; but Daoism is beyond such contingencies. (2004, 66).

A second relevant Western thinker is the evolutionist theologian Pierre Teilhard de Chardin, who much in line with Daoist thought argues for a theology of *fieri,* one centered on becoming rather than on *esse* or being. For him a metaphysics of being prevents the very possibility of any onto-cosmic betterment because Being is conceived as already complete, and does not allow, therefore, for any participation or contribution on the part of mankind and history.

Then again, D. T. Suzuki is the most celebrated early Asian expositor of Buddhist philosophy in the West. In particular, he devoted several seminal books to Zen, including *Zen and Japanese Culture,* published first in 1938. He studied Zen under Master Soen Shaku (1860-1919), who recommended him to the German scholar Paul Carus (1852-1919) for a collaborative project of English translation of Eastern classics. Thus, Suzuki published an English translation of the *Daode jing* in 1913. Although none of his writings are specifically devoted to Daoism, many include references to it while highlighting the specifically shamanic characters of the tradition. Moreover, Suzuki contemplates Daoism from the point of view of its convergences with, and contributions to, Zen Buddhism.

The French metaphysician, the Swiss theologian, and the Japanese philosopher belong to the same generation of thinkers. Born in the late 19th century, the age of the advent of progressive positivism, they died after World War II, which crystallized a global trauma and fostered a pessimistic and apocalyptic vision of time and history.

Guénon's Affinities with Daoism

Guénon's interpretation of time is inspired by Hindu eschatology through the doctrine of the cycles or *yugas*. He sees manifestation as entailing a gradually widening separation from the Divine Principle, hence fall and decadence. As time unfolds, the imperfections and oppositions grow, until they finally lead to disintegration. This is a law of metaphysical entropy, which means that the highest degree of reality lies in the primordial because it is still "fresh from" the Source. Guénon's onto-cosmogenesis reveals affinities with the metaphysics of Laozi and Zhuangzi. Let us consider, for instance, chapter 25 of the *Daode jing*:

> There was something undefined and complete,
> Coming into existence before heaven and earth.
> How still it was and formless,
> Standing alone and undergoing no change,
> Reaching everywhere and in no danger (of being exhausted)!
> It may be regarded as the Mother of all things. . . .
>
> Great, it passes on (in constant flow).
> Passing on, it becomes remote.
> Having become remote, it returns. . . .
>
> Man takes his law from the earth;
> The earth takes its law from heaven;
> Heaven takes its law from Dao.
> The law of Dao is being what it is. (Legge 1891, 67)

In its primordial and formless aspect, Dao is identifiable with original chaos, Hundun 混沌. It comes first in time, or rather first before time, and everything flows from it. Legge's "undefined and complete" translates the two characters *hun* 混 and *cheng* 成. *Hun* can mean "including everything" in a state of indifferentiation while *cheng*, which literally means "to complete," implies a reality that is fully formed and lacks nothing.

The second character used in the chapter to refer to chaos, right before *huncheng*, is 物, which means "thing, that, is substance or a form of being.[1] The Absolute manifests itself in the difference between something and mere absence or unreality. It is the independent Ground of everything

[1] Compare Frithjof Schuon's characterization of the Ultimate as Absolute, Infinite, and Perfect (2000, 26).

that distinguishes something from nothing. As for *hun*, it is what Guénon would call the All-Possibility, another term for the Infinite. It is "formless" (Mitchell 1988) because it precedes the production of any form. By contrast, *cheng* refers to "something perfect," whole or complete.

The four elements follow the creative energy of Dao, from the original mystery to the universe, then to earth and to humankind. In reverse, each needs to conform to the one preceding in the chain of being, that is "take his law." The word for "law" (*fa* 法) connotes imitation, reproduction of patterns but also abiding by laws. The need to "follow" the law of the higher-level entity signifies an ontological dependence. In this respect, the last ring, the human state, differs since it is characterized by its capacity not to follow the law it receives from the earth. While an ontological fall manifests an increasing contingency on the part of beings, an epistemological and spiritual decay results from human straying from the law flowing from Dao. Legge denotes this straying through the word "remote" (*yuan* 遠). The character suggests a going away, a distance, but it also implies a final return (*fan* 返), both containing the "walk" radical.

Another point of comparison with Guénon's onto-cosmogony appears in chapter 42. As James Legge renders it,

> Dao produced the One;
> The One produced the Two;
> The Two produced the Three;
> The Three produced the ten thousand things.
> All things leave behind them the Obscurity [from where they come],
> And go forward to embrace the Brightness [into which they emerge],
> While they are harmonized by the Breath of Vacancy. (1891, 85)

The two last words translate *chongqi* 沖氣, combining the term for "bland," "empty," "vacant," and "thrust" with that for "breath," more fundamentally indicating vital energy or the life force. Obscurity refers to yin, while Brightness translates yang. The two connote non-manifestation and manifestation, shadow and light as evident on north- and south-facing slopes of hills. While yin and yang are principles of multiplicity, they are so only in relation to a third element which is both principle of energy and void.

Thus, Daoists associated the number three with the production of the ten thousand things as the alternating and productive aspect of yin-yang can only unfold in the Breath of Vacancy. Three is the number of becoming, one and two being in different ways more static. One is literally not a number—it is transcendent—while two may refer to yin and yang, as in the duality of heaven and earth, but it may also correspond more specifically to yin and earth in so far as it is the first even number, and directly related to four, the number of the earth and the cardinal points. Three, as the first odd number will be paired, by contrast, with yang and heaven.

Moreover, the cyclical dimension of time is symbolically akin to the number three, associated to the unfolding sequence of the ten thousand things. Can this connection with time be contemplated as bound with the metaphysical necessity of fall and decay?

A Daoist Fall?

Three, moreover, can also correlate with space. The Chinese triad of heaven, humankind and earth is one of the most significant examples of such a spatial ternary. In *The Great Triad*, Guénon delves into the question of the transposition of this spatial triad in the domain of time. In this scheme of things, the present as the intermediary between past and future is analogically akin to the position of humankind in the Great Triad, located between heaven and earth.

The instantaneity of the present, then, manifests an analogy with human free will, its ability to emancipate itself from the determinations of earth and heaven. Free choice can only be exercised in the present, while past and future pertain to necessity, albeit in different ways. Guénon relates the former to Destiny and the latter to Providence, which are respectively "causal" and "final" determinations. In Hindu and Buddhist terms, they would refer to the karmic chain or warp on the one hand and the weft of divine or bodhisattva grace on the other hand.

Human freedom must also be correlated with *qi* 氣, the "space of energy" where the combination of yin and yang takes place. As Guénon says, "The ten thousand things are produced by Taiyi 太一 [lit., the Great One] and modified by yin and yang" (2004, 35). That is, heaven and earth are one in the Principle of the Great One, the root state of yin and yang, which Daoists, based on the *Yijing*, often also called Taiji 太極, the Great Ultimate. Their polarization makes it possible for production to occur "in the 'interval' between them." The *Daode jing* proclaims,

> The space between heaven and earth,
> Isn't it like a bellows?
> Empty, yet never bent;
> Active, yet reaching even further. (ch. 5; Kohn 1993, 14)

Likewise, humankind's free will is a place of interaction between heaven and earth, along the axis that joins the transcendent principle or

Great Ultimate to all other domains of existence. The human state is perfected in the union of heaven and earth.

While the *Daode jing* envisions the production of the ten thousand things as the manifestation of the creative power of Dao, it makes no explicit mention of a law of ontological entropy. One of the reasons why Daoist texts do not dwell on the reality of metaphysical degeneracy may lie in their contemplation of Dao from an immanent point of view, as *natura naturans*. Dao is an eternal principle that unceasingly produces the myriad beings and forms: therefore, its activity knows no fall and no diminution. The great Daoist classics emphasize immanence and production rather than the distance and fragmentation that they entail as an "unintended consequence."

Thus, ontological degeneration does not appear in the foreground of Daoism, but is entailed by the consideration of its spiritual and ethical dimensions. In fact, the immanence of Dao paradoxically accounts for the possibility of its loss. Its presence is sometimes characterized as small and humble, non-acting, subtle, and imperceptible. As the *Daode jing* says:

> The Great Dao flows everywhere,
> (Like a flood) it may go left or right.
> The myriad things derive their life from it,
> And it does not deny them.
>
> When its work is accomplished,
> It does not take possession.
> It clothes and feeds the myriad things,
> Yet does not claim them as its own.
>
> Often (regarded) without mind or passion,
> It may be considered small. (ch. 34; Lin 1948)

The humility of Dao—"it can be considered humble" (McDonald 2017, 34)—means that it does not impose itself on humankind or, as James Legge says, "It does claim to be their lord." This means in return that all beings are free to divert themselves from it, even to obstruct its channels of communication and flow.

This relates to the very opening lines of the *Daode jing*: "The Dao that can be told is not the eternal Dao" (ch. 1). Particularity and multiplicity are bound to naming, while Unity at the root of all existence is unfathomable and unnamable. Beyond naming, desire exteriorizes people's perception of reality and prevents them from attaining the mystery of things. The chapter lays out the very principle of the fall in the sense that it offers an understanding of how the human connection to Dao can be lost. Dao itself is not susceptible to change, nor liable to fall, but there can be a disconnection from it that entails a lessening of its power of inner transformation. Thus, in the same chapter, the distinction between the "wonder" (*miao* 妙) of Dao and its "outcome" (*jiao* 徼). As the text has,

> Always remain free from desires—
> And you can see its wonder.
> Always cherish desires—
> And you can only observe its outcome. (ch. 1; Kohn 1993, 13)

Wonder and outcome refer to the kernel and the shell of Dao. Wonder connotes subtlety and mystery, referring to the hidden essence. Outcome, by contrast, connotes form, extremity, or limit. The first is like the unfathomable beginning, while the second is the perceptible end.

Laozi implies that the outcome entails less being and energy than the source, but his way of approaching this difference is mystical rather than metaphysical. The key term here is *yu* 欲, which connotes desire and attachment. While the Daoist perfected human being abandons desires so as to perceive the mystery that lies beyond the limitations of objects, the man of desire remains bound by the horizon of peripheral forms and has therefore no knowledge of the essence of things.

The principle of metaphysical decay, in other words, is closely involved in the way that humans alert to the dangers of being separated from original Unity, veiled from it by petty and narrow concerns and distinctions, by inordinate desires for outer phenomena. The text notes that the origin of the universe is like a mother, from whom sons are born. The sons' knowledge forms part of the existential lot of humanity, but people should "keep to the mother" if they wish to be "preserved from harm.

> There was a beginning of the universe,
> Which may be regarded as the mother of the universe.
> From the mother, we may know her sons.
> After knowing the sons, keep to the mother.
> Thus one's whole life may be preserved from harm. (ch. 52; Lin 1948)

A Daoist View of Progress

From another point of view, Daoist thought may yet also seem to propose a progressivist consideration of reality through the concept of the transformation of things. Daoism is not given to highlight the permanence of essences, but rather emphasizes the transmuting presence of the creative principle that flows through everything. Is this akin to modern views of becoming and a progressivist *Weltanschauung*?

There is probably no more strikingly representative of the latter as evolutionist theologian Pierre Teilhard de Chardin, who argues for a theology of *fieri* (becoming) rather than *esse* (being). For him a metaphysics of being prevents the very possibility of any onto-cosmic betterment because being is conceived as already complete, and does not allow any participation or contribution on the part of humankind and history.

By contrast, Telihard sees the Incarnation as marking the entrance into a cycle of realization of the Spirit in and through the cosmos. Through human thinking and realizations, history is given a revelatory meaning and culminates in the Omega Point of a cosmic apogee of Divine Consciousness. Thus, there is no real Absolute without the perfecting culmination of the universe through human and cosmic transformations.

Against a traditional metaphysics of Unity akin to Guénon's, Teilhard argues for a "metaphysics of Union." What does it mean?

> In the metaphysics of *esse*, pure act, once posited, monopolizes all that is absolute and necessary in being; and, no matter what one does, nothing can then justify the existence of participated being.
>
> In a metaphysics of union, on the other hand, we can see that, when once immanent divine unity is complete, a degree of absolute unification is still possible: that which would restore to the divine center an 'antipodial' aureole of pure multiplicity. . . . The created, which is 'useless', superfluous, on the plane of being, becomes essential on the plane of union." (1974, 178)

In this view, Creation is "antipodial" because it stands at the antipode of Divine Unity, being utterly independent from it. But it is also like an "aureole" in the way it endows the Divine Reality with a sort of ontological glow. The world is the aureole of God, and as such it has a necessary function in the advent of the Truth. For Teilhard, the Absolute is in some sense a reality in progress, a reality that is fulfilled in the end as an ultimate Revelation: the so-called Omega Point. In other words, that which is not God needs be taken in some way as a kind of second absolute, without which there would be neither striving toward Union nor Union itself.

Now, notwithstanding its being the immanent principle of constant transformations, Dao is also transcendent in and of itself. The idea of Dao being augmented, perfected, or fulfilled in any way is profoundly foreign to the Daoist philosophical outlook. Dao does not lack anything, and becoming or transformation is not a process through which its being would be fulfilled.

Moreover, the very idea of a union with Dao is strictly speaking ill-sounding in Daoist metaphysics, since union presupposes a duality that the uniticity of Dao precludes. Besides, even from the point of view of onto-cosmic reality, the alternation of yin and yang is not directional: it does not point toward an apogee, an Omega Point. This cosmic alternation revolves around a motionless center. It does not move away from this center in any ascending manner, but punctuates the rhythm of cosmic existence in a sequence of manifestation and return.

If there is a way, however, in which progress can be said to be part of the Daoist *Weltanschauung*, it is in the principle of a return to Dao. This return may be passive and necessary or active and merely possible: everything flows back ultimately into Dao, while the sage returns purposefully to it. The most conscious and effective return is spiritually driven: it results from an inner discipline that aligns with Dao.[2] But far from being a human production, this spiritual process entails a lowering, a diminution, a non-doing (*wuwei* 無為) one that gives free passage to the transforming Dao through "fasting of the mind."[3] Daoism does not magnify human enterprises but favors a minimalist proximity to the rhythms of nature. It could not be more remote from the idea of counting human ideas, sciences, and techniques as integral elements in a development of the Absolute. The Daoist sage is simply a kind of know-nothing who knows everything.

Even the collective dimension of a civilizational progress or apogee does not readily appear as part and parcel of the message of Daoist classics. Consider, for instance, how—far from being a contribution to a progressing civilization—human skills are contemplated by Daoist sages as an attunement to Dao, which amounts to a quasi-vanishing of the human. Thus, Zhuangzi's Cook Ding is not an active participant in the progress of human technology: he simply follows in the footsteps of Dao and its rhythms, his own selfhood disappearing in the very process of the unfolding of nature. Human skills are not inventions, rather, they flow from a unison with Dao.

---

[2] "The only real progress that can be made is in inner space, in the attainment of enlightenment which releases from the concepts and bondage of both space and time." (Cooper 2010, 37).

[3] Yan Hui asked, "What do you mean by the fasting of the mind?" Confucius replied, "Bring all the activity of the mind to a point of union. Do not listen with your ears, but listen with the mind (thus concentrated). (Then proceed further and) stop listening with the mind: listen with the spirit [vital energy]" (*Zhuangzi* 4; Izutsu 1983, 343).

Human Straying from Dao

While wise human beings do not superimpose anything on reality and do not try to perfect it, their unwise fellows can precipitate indeed a fall from it with the best of intentions and most active doing. Here is a typical account of the Daoist concept of fall:

> When the great Dao declined,
> The doctrine of humanity and righteousness arose.
> When knowledge and wisdom appeared,
> There emerged great hypocrisy.
> When the six family relationships are not in harmony,
> There will be the advocacy of filial piety
> And deep love to children.
> When a country is in disorder,
> There will be the praise of loyal ministers. (ch. 18; Chan 1963, 131)

Therefore, whatever increasingly lacks on the highest level is partially compensated on a lower one, always at the cost of a greater proximity to the principle. Goodness and piety are ethical and religious mediations only necessary because the immanence of Dao is neither recognized nor reached anymore. The "men of old" did not need to be "good" and "pious" since their very being was in agreement, indeed in conjunction, with Dao. In fact, all human forms and productions are just poor substitutes for Dao and its natural manifestations.

Zhuangzi is even more explicit in describing human straying:

> The men of old, their knowledge had arrived at something: at what had it arrived? There were some who thought there had not yet begun to be things—the utmost, the exhaustive, there is no more to add. The next thought there were things but there had not yet begun to be borders. The next thought there were borders to them but there had not yet begun to be 'That's it, that's not.'
>
> The lighting up of 'That's it, that's not' is the reason why Dao is flawed. The reason why the Way is flawed is the reason why love becomes complete. Is anything really complete or flawed? Or is nothing really complete or flawed? (ch. 2)

The first rank of men of old thought that things "had not yet begun to be." This is the pure non-duality of Dao, nothing "is" in the sense that there is only being, a far cry from Teilhard's ontology of *fieri* and Union. The second stage appears when human consciousness witnesses things, albeit without "borders" between them. Multiplicity is still lived within unity, whereas at the precious stage pure unity was lived without any awareness of multiplicity. The third stage entails the perception of borders. The pristine unity of things is now affected by determinations.

Here borders are ontological outlines. Humans are aware of them, but they do not yet superimpose upon them conceptual and ethical distinctions. The fourth stage involves distinctive knowledge *qua* rational distinctions and value judgment: "that's it, that's not," or right and wrong. At this stage human epistemology divides reality along mental and conventional lines. It is precisely at this moment that "Dao is flawed."

The flow of Dao is obstructed, as it were, by the *de facto* absolutization of human and subjective distinctions and differences. The subjective perception has overwhelmed objectivity, and this substitution is clearly signaled by the recourse to love, a form of benevolence that is here less the inherent offspring of human nature than a moral and sentimental imperative made necessary by the distinctions and disorders introduced by human categories. It is an all too human way to try to make up extrinsically for the metaphysical sense of unity that has been lost.

The final question—"Is anything really complete or flawed? Or is nothing really complete or flawed?"—is reminiscent, in typical Daoist fashion and lest we start overly dramatizing the cosmic downfall, of the fact that all these distinctions have no true reality in the sense that they do not touch upon Dao itself. So the Daoist view of history emphasizes the reality of a human moving away from the Principle of being and consciousness that echoes some of Guénon's views on human history.

Manifestation and History

Guénon's consideration of the descending motion of the cycle and the multiple disorders it entails raises two difficulties. First is the question whether such a negative view of manifestation, and the fall it speaks of, ignores the fact that manifestation as such is good or that it is fundamentally none other than Dao. Second is the issue whether there is any form of freedom and responsibility left for humans in this seemingly inexorable view of the fall. The latter relates directly to Daoist historical accounts, since they raise the possibility for humankind to either follow in the wake of Dao or lose touch with it.

Guénon's works address the first question in his assertion that manifestation, notwithstanding its negative dimension of separation from Principle, actualizes realities, phenomena, and human productions that would remain un-manifest without it. In this respect, it allows for the actualization of positive possibilities. In other words, what is loss and distance from a certain point of view is development or growth from another one, and indeed even "partial disorders . . . contribute . . . to the total order."[4] Still, the aspect of downfall is more significant than the aspect of growth, because loss betrays the perfection of the Origin, whereas the unfolding of possibilities can only manifest perfection indirectly, and therefore perforce imperfectly, if one may say so.

However, the relative positivity of manifestation is also perceptible in terms of the return to Dao it affords. This is particularly akin to the function of virtue (*de* 德). Thus, we read in the *Daoti lun* 道體論 (On the Embodiment of Dao), an 8th-century exegesis of the *Daode jing*.

> Dao is all-pervasive; it transforms all from the beginning.
> Virtue arises in its following; it completes all beings to their end.
> They thus appear in birth and the completion of life.
> In the world, they have two different names,
> Fulfilling their activities, they return to the same ancestral ground.
> They are two and yet always one. (Kohn 1993, 19)

Virtue, therefore, is in a sense the redeeming essence of manifestation.

As regards the second question, Guénon clearly sees human responsibility in forgetting Dao, which comes acutely to the fore with the advent of what he calls "the modern world." For him, something fundamental happened in Europe by the end of the Middle Ages, crystallizing more evidently with the Renaissance: a theocentric and qualitative universe was gradually replaced by an anthropocentric and quantitative vision. In order to understand Guénon's critique of the modern world, two phases are essential that he sees as determining the final stages of the cycle.

There is, first of all, the emergence of what he calls the "anti-tradition" and then, at a second stage, the advent of what he names "counter-tradition." When using the term Tradition, Guénon does not refer to any particular religious tradition or even less to specific customs or conventions, the latter being merely human constructs or historical accretions. For him, Tradition is transcendent or divine, and it constitutes the universal heritage of all humankind throughout the ages. All religious traditions are like branches of the tree of Tradition.

By contrast, an anti-traditional stamp characterizes materialism, as well as all secularizing ideas and trends. Guénon sees those as producing a "solidification" of the periphery of the universe. The way human beings

---

4 "All partial disorders, even when they appear in a certain sense to be the supreme disorder, must nonetheless necessarily contribute in some way to the total order" (Guénon 2000, 265).

think and feel has an effect on the cosmos, while the latter determines the human apprehension of the objective field.

The cosmic solidification is a particular mode of active participation in the metaphysical law of entropy. In another sense, however, the human self is a prime locus of consciousness and freedom. As he says,

> The materialist conception, once it has been formed and spread abroad in one way or another, can only serve further to reinforce the very 'solidification' of the world that in the first place made it possible. (2000, 117)

This view is akin to the Daoist critique of the hardening characterizing the loss of Dao. It corresponds to Zhuangzi's "fixed mind" in contrast to the fluidity of the mind of the sage (Izutsu 1983, 425). A growing deafness and hardness of humankind results in an increasingly deafening silence of the gods. Thus, the natural order can become the prey of human industrial activism and exploitative desecration, the very opposite of non-doing.

Guénon does not see anti-traditional action as the last word, however. It is followed by a moment he characterizes as "counter-traditional." For him, while "anti-tradition" is a "deviation," "counter-tradition" is akin to a "subversion." The former is gradual whereas the latter is sudden, when the scales are abruptly tilted against Tradition. He notes,

> "[The counter-tradition] will in the end contrive to 'exteriorize', if that is the right word, something that will be as it were the counterpart of a true tradition, at least as completely and as exactly as it can be so within the limitations necessarily inherent in all possible counterfeits as such" (2000, 261).

While anti-tradition rejects Tradition, the "counter-traditional" moment is one of parody of some of its components, for the counter-tradition also aims at building a new world that has all the appearances of a religion and a spirituality. As a matter of fact, its character of parody enables the counter-tradition to make use—especially in its early phases—of "elements authentically traditional in origin, perverted from their true meaning, and then to some extent brought into the service of error" (2000, 269).

This is the case with regard to spiritual elements from Asian traditions, including Daoism, which are severed from their traditional contexts and amalgamated with counter-traditional ideas and practices. The relative fluidity of Eastern metaphysical teachings (including Daoism) makes them particularly susceptible to such reductive and parodic misuses. All in all, Guénon sees counter-tradition as an instrument of "dissolution" that follows the anti-traditional "solidification."

While counter-tradition is the agent of final dissolution, the latter is not all negative. In fact, Guénon's view is that the final stages of the dissolution see a kind of "pulverization" and "volatilization" of reality, which, beyond its destructive aspect, opens the way to "crystallization" and "sublimation" (Guénon 2000, 168). In his own words,

> Bodies can then no longer persist as such, but are dissolved into a sort of "atomic" dust without cohesion; it would therefore be possible to speak of a real "pulverization" of the world, and such is evidently one of the possible forms of cyclic dissolution. (2000, 167)

These two processes contribute to a final "rectification,"[5] a sudden return to Reality that is not strictly inscribed in time, being like an instantaneous reversal of polarities. In this time beyond time, the line of demarcation between two cycles cannot be absolute, and it must be so since, in spite of gaps and chasms, continuity runs everywhere through the cosmos like a common thread. In this regard, Guénon refers at times to "seeds" present in one cycle that will bear fruits in the next one, although the restoration brought about by these seeds cannot take place without "the immediate intervention of a transcendent principle" which must "fix" the positive germs of the future cycle. The way in which this "fixing" takes place is envisaged by Guénon in light of the alchemical principles of *solve* and *coagula*.

In their negative characters, these two principles correspond to "precipitation" and the "return to the indistinction of chaos." What is precipitated is the *caput mortuum*, the worthless residues that ultimately go back to a state of chaos. However, what truly remains are the elements crystallized in, and reintegrated by, the alchemical process. Thus, positively, *coagula* and *solve* correspond to the alchemical principles of crystallization and sublimation. That which is crystallized, solidified in a positive sense, is subsequently sublimated and reintegrated onto a higher ontological level, dissolved in an analogous positive sense.

Guénon asserts that negativity and subversion are always temporary and indeed illusory. In the last analysis, the source of any error and evil is to be found in metaphysical dualism, since the latter divides and opposes, whereas Reality is essentially one. Beyond and even within multiplicity

---

[5] "At the very moment when it seems most complete it will be destroyed by the action of spiritual influences which will intervene at that point to prepare for the final 'rectification'" (Guénon 2000, 261).

and disorder lies the unity of the Source. In Guénon's eschatological scheme, this is expressed by the fact that the passage from one cycle to the next, from an end to a new beginning, must include both discontinuity—there is no new life without *solve*—and continuity—any *solve* entails *coagula*.

Yin-Yang and the Alchemy of Time

Another important aspect of these considerations is the question of the sequencing of cosmic qualities:

> In the unfolding of the cosmogonic process, darkness—equated with chaos—is "in the beginning," whereas the light that brings order into this chaos and out of it produces the Cosmos comes "after the darkness." This amounts yet again to saying that, in this particular context, yin effectively comes before yang. (Guénon 2004, 32)

Chaos is both "formless" and "complete." Formless, it needs the "light" of form and order to produce the cosmos; complete, it is the essential principle of the ten thousand things. Moreover, there is need to introduce another distinction, this one vertical and not horizontally sequential, between the metaphysical yin and the cosmogonic yin. In this regard, it can be said that the primordial chaos is formless when contemplated as the Undifferentiated All-Possibility. Several passages from Laozi identify it as a feminine Reality, such as the valley spirit, the mother, or the mysterious female.

Inasmuch as it lies beyond cosmology, this Abyssal Reality cannot be properly considered as either yin or yang. There is a sense, however, in which a symbolically yin marker can be applied to the Principle of everything, given that its primordial reality is non-manifested and therefore mysteriously inward and dark (xu*an* 玄), a term that appears, for example, in chapter 6 of the *Daode jing*.

> The spirit of the valley dies not, aye the same
> The female mystery thus do we name.
> Its gate, from which at first they issued forth
> Is called the root from which grew heaven and earth. (Legge 1891, 51)

When considered from a strictly cosmogonic point of view, the chaos Guénon refers to may correspond to pure yin. In this case, however, the twofold characterization of Chaos as formless and complete does not apply, since the cosmogonic chaos cannot be said to be complete as it needs its yang productive complement in the form of "the light that brings order into this Chaos and out of it produces the Cosmos." Moreover, in regard to time, a consideration of Chaos as formlessness implies that the cosmogonic process can be considered as a kind of progress in the sense of unfolding. From the metaphysical point of view of its completeness, by contrast, everything coming from it can only be a loss of being, i. e., a sort of decay and entropy.

Guénon's correspondence between the two pairs yin-yang and *solve-coagula*, the first being cosmogonic and the second alchemical, is based on an association between heaven and yang, on the one hand, and earth and yin, on the other. Yin here implies a sort of productive condensation and solidification, which in Chinese symbolism is associated with moisture. Yang, in contrast, corresponds to air and dryness: it is seen as a principle of dissolution and therefore a mode of return to non-manifestation. It is also associated with *qi* (Robinet 1993, 83). That which is heavy and turbid becomes earth; that which is light and limpid becomes heaven (Izutsu 1983, 305).

In *The Great Triad*, Guénon explains that the order of *solve* and *coagula* depends on the vantage point adopted (2004, 44), that is, whether one starts from a state of non-manifestation or a state of manifestation. In the first case *coagula* comes first, as productive condensation, and *solve* comes in second. However, when manifestation is the starting point, there is first *solve*, then *coagula*.

The "manifestation" of the modern world corresponds to *coagula*, a solidification. It is the result of a process of prior dissolution, identified by Guénon with the Renaissance. This period, by a shift away from theocentric to humanistic principles, paved the way for a new world founded on the negation of metaphysical principles. Once this materialized outlook solidified, it became impervious to the supernatural influences external to its definition of reality.

The positive side of this solidification lies in its limitations; it creates an impression of ordinary security: although illusory, it defines as it were a particular "reality." This false security dissolves with the entrance into what could be called post-modernity. The final dissolution culminates with the conjunction of a crystallization and a sublimation: that which is fixed is sublimated. The best of the earth, as a crystal, is sublimated into a new beginning, and this means passing from an old to a new state of manifestation, the beginning of a new cycle.

Mountains, Rivers and the Search for the Ox

As a transition to Suzuki's world of Zen, it is significant that the simultaneity of *solve* and *coagula* presents a profound similarity with the Zen parable of the three stages in the spiritual perception of reality, in the symbolic form of "mountains" and "rivers." In the beginning, mountains are mountains and rivers are rivers, then mountains are not mountains and rivers are not rivers, and finally mountains are mountains and rivers are rivers again (Heine 2000, 301).

These three moments correspond to the conventional view, to the reduction of everything to the no-thingness of *sūnyatā*, and finally to the restoration of everything to its Buddha-nature. Toshihiko Izutsu has noted that the second and third are but two faces of a same reality: "At the last stage, 'A is A' is but an abbreviated expression standing for 'A is non-A'; therefore, it is A" (2001, 29). The emptying of emptiness is simultaneous with the realization of emptiness. If stage two is to be what *it is*, stage three "establishes itself at the same time." The second and third stages are coincidental, and their sequencing can only be logical, not existential.

Such paradox of time and transcendence is also found in the Buddhist parable of the search for the ox. It displays ten stages of enlightenment, from an initial search to supreme awakening. In his commentary of this visual parable, Suzuki refers to a Zen classic from the 12th century and argues that it is paradoxical to speak of stages of enlightenment, since enlightenment in Zen is described more as abrupt or sudden than as gradual.

There is indeed a paradox of discontinuity in continuity in the first picture, depicting the man departing in search of the ox. Indeed, "the beast has never gone astray, so what is the use of searching for him?" (Suzuki 2015, 153). The very notion that we must look for enlightenment means that "as long as the man is conscious of his 'Self' in connection with the prize, there is the dualistic separation of the possessor and the possessed" (2015, 151).

Similarly, even though we look for the ox everywhere, its "horns or rather nose is said to reach the heavens and there is nothing that can hide him. It is we who shut our own eyes and pitifully bemoan that we cannot see anything" (2015, 151). The problem is, therefore, spiritual or subjective, a problem that the various stages of the training dramatize. The supreme paradox is that the various stages are all "for nothing," since all traveling is ultimately the result of an illusion. There is no time, no fall, no redemption.

With regard to time, the eighth and ninth illustrations are most significant. The plain circle in the eighth picture symbolizes emptiness, an emptiness that is all-creative and inexhaustible. Zen teaches that, when everything has seemingly disappeared into emptiness, the mountains are not mountains and the rivers are not rivers. The realities of our ordinary perception lose their hold, as their nothingness is revealed. Flowing from the eighth, the ninth picture shows nature as it is, not as we imagine it or distort it, "mountains are mountains and rivers are rivers."

This third approach to time means transcending it, as in the unity with Dao, which is atemporal, infinite, and eternal. While the cosmic wheel turns in time, its axle dwells beyond, although each point of the periphery is related to the motionless mover. Daoism, therefore, is both fundamentally eternalist and keenly sensitive to the ways Dao plays out in time through the ten thousand things. The unity of the whole is perceived in time and beyond time: even decay connotes return.

By contrast, it could be argued that Zen Buddhism grasps reality in the space of an instant. Suzuki's meditations on *satori* provide us with insights into this vision of time. Zen perception is instantaneous and discontinuous. Time as fall and time as progress have no reality. In Zen enlightenment, eternity has "no time," and the relationship between what precedes and what follows is mere appearance. What truly is can never follow what is not.

Zen is essentially trans-temporal and with it time changes into space, albeit not in the way of a Guénonian "catastrophe," but rather like an "anastrophe," a kind of "falling upward." In the flash of *satori*, everything is given together and at once. The emphasis is not on transformation, but on instantaneity. There is a shift from time as entropy or progress to time—or rather no-time—as the instant of eternity.

One must be careful, though, not to understand this emphasis on instantaneity as a negation of time. As is well known, Zen lays emphasis on the task at hand, the practical. In this regard, one of Suzuki's most striking statements is the following:

> To Zen, time and eternity are one. This is open to misinterpretation, as most people interpret Zen as annihilating time and putting in its place eternity, which to them means a state of absolute quietness or doing-nothing-ness. They forget that if time is eternity, eternity is time. . . .
>
> Zen has never espoused the cause of doing-nothing-ness; eternity is our everyday experience in this world of sense-and-intellect, for there is no eternity outside this time-conditionedness. An "objective" eternity has no meaning for us; there is no "quiet place" outside of time where we could experience it. (1956, 266)

Buddhism, weary as it is of metaphysical objectifications, leads one to parry a kindred substantialization of eternity. When stating that "there is no eternity outside this time-conditionedness," Suzuki challenges conventional reason in typical Zen fashion.

What can be the meaning of eternity if it is conditioned by time? Suzuki does not state that eternity is conditioned by time, he simply affirms that it cannot be experienced outside of time, but on the contrary must be actualized within and through time. This is not a doctrinal statement: the stress lies on a spirituality that does not separate eternity from time because it teaches us how to realize it practically in the most attentive presence to our timely condition.

In other words, such a view of eternity is akin to *koans* such as "What is the Buddha? Three pounds of flax!" In relating this *koan*, Suzuki shifts the attention of his reader from the temptation to understand it in metaphysical, pantheistic, terms, to a spiritually practical insight into the "inmost recess of . . . consciousness" (1964, 79).

Time and Enlightenment

The paradoxical idea that eternity is time and time eternity is shared by Zen and Daoism by virtue of their non-dualism. However, the concept of eternity appears in Daoism with a different emphasis. It stresses the perpetual non-active action of Dao ever renewing the ten thousand things. It sees the instant as transformative. While Zen perceives multiplicity within the unity of the Buddha-nature, as in a flash, Daoist wisdom aims at tapping into unity through and within the ten thousand things that flow from and with Dao.

Spiritually, it is significant that Daoist "fasting of the mind" and Zen *satori* are characterized by different moods. The former is a forgetting of everything that makes it possible to connect inwardly with Dao. Thus, Laozi's closing of apertures:

> Close the mouth.
> Shut the doors [of cunning and desires].
> And to the end of life
> There will be [peace] without toil. (ch. 52; Chan 1963, 52),

This is perhaps why Daoist meditation is less intent on visual contemplation: "Even without peeping out of the windows, one can see the working of heaven" (Izutsu 1983, 338). The time of the "working of heaven" is none other than eternal Dao; outer and inner time are not different. "He [the sage] goes on producing within his 'interior' the 'time' [of the world]" (1983, 346). This is because the Daoist sage is at one with the alternation of yin and yang. The perfect human being brings everything back to Dao by being a pure mirror of the transformation of things. Beyond, however, there is not even any sense of distinction, like the "men of old" for whom things had not begun yet.

Although the sitting meditation of *zazen* is also a temporary shutting out of external perceptions, *satori* is like a glimpse into totality and a definitive affirmation; it is a maximal opening. Suzuki sees it as affirmative and accepting: "Though the *satori* experience is sometimes expressed in negative terms, it is essentially an affirmative attitude towards all things that exist; it accepts them as they come along regardless of their moral values" (1956, 103).

Another connected aspect of the Zen experience is its momentariness: "*Satori* comes upon one abruptly and is a momentary experience. In fact, if it is not abrupt and momentary, it is not *satori*" (1956, 103). It coincides with an instantaneous perception that "rivers are rivers and mountains are mountains." In Daoism, by contrast, the end point might very well be that "rivers are mountains and mountains are rivers." This is because the main focus is on transformation. This also accounts for the fact that eternity and perpetuity, or eternity and immortality, are difficult, if not impossible, to distinguish from each other.

The contrasts we have sketched should not be over-stated, however. Kenneth Inada cautions:

> These two methods are not really contradictory since Zen, for example, incorporates the quietistic nature in its meditative process. There is actually no difference in the Daoist "forgetting himself" and the Zennist concept of losing his self. Any devotee, either Daoist or Zennist, may spend hours "honing up" for the final grasp of reality, but he must not waste his time in futile "brick grinding" to produce a mirror, or in squeamish rituals upholding Confucian virtues. (1988, 62)[6]

"Wasting time" with "doings" amounts to not understanding that time is regained only by transcending it, by "idling away one's time" (Izutsu 1983, 435) with Zhuangzi by the side of the tree of nondoing.

> You have this big tree and you're distressed because it's useless. Why don't you plant it in Not-Even-Anything Village, or the field of Broad-and-Boundless, relax and do nothing by its side, or lie down for a free and easy

[6] "Both Zen and Daoism have already conquered the minds of Asians (and many non-Asians, too, for that matter) by simply rendering clear 'some eternal greatness incarnate in the passage of temporal fact'" (Inada 1988, 52).

sleep under it? Axes will never shorten its life, nothing can ever harm it. (Watson 2013, 6)

Bibliography

Bidlack, Bede Benjamin. 2015. *In Good Company: The Body and Divinization in Pierre Teilhard de Chardin, SJ and Daoist Xiao Yingsou*. Leiden: Brill.

Chan, Wing-Tsit. 1963. *The Way of Lao Tzu: Tao te ching*. Upper Saddle River, NJ: Prentice-Hall.

Cooper, Jean C. 2010. *An Illustrated Introduction to Daoism: The Wisdom of the Sages*. Bloomington, Indiana: World Wisdom.

Guénon, René. 2000. *The Reign of Quantity and the Signs of the Times*. New Delhi: Munshiram Manoharlal.

_____. 2004a. *The Great Triad*. Hillsdale, NY: Sophia Perennis.

_____. 2004b. *Insights into Islamic Esoterism and Taoism*. Hillsdale, NY: Sophia Perennis.

Heine, Steven. 2000. *The Kōan: Texts and Contexts in Zen Buddhism*. Oxford: Oxford University Press.

Inada, Kenneth. 1988. "Zen and Daoism: Common and Uncommon Grounds of Discourse." *Journal of Chinese Philosophy* 15:51-65.

Izutsu, Toshihiko. 1983. *Sufism and Taoism: A Comparative Study of Key Philosophical Concepts*. Berkeley: University of California Press.

_____. 2001. *Toward a Philosophy of Zen Buddhism*. Boston: Shambala.

Kohn, Livia. 1993. *The Taoist Experience*. Albany: New York: State University of New York Press.

Legge, James. 1891. *The Sacred Book of China: The Texts of Taoism, Part I*. Oxford: Clarendon Press.

Lin, Yutang. 1948. *The Wisdom of Laotse*. New York: Random House.

McDonald, J.H. 2017. *Tao Te Ching: An Insightful and Modern Translation*. Troy, NH: Qigong Vacation.

Mitchell, Stephen. 1988. *Tao Te Ching: A New English Version*. New York: Harper-Perennial.

Robinet, Isabelle. 1993. *Taoist Meditation: The Mao-shan Tradition of Great Purity*. Albany: State University of New York Press.

Schuon, Frithjof. 2000. *Summary of Metaphysics and Esoterism*. Bloomington: World Wisdom.

Suzuki, D. T. 1956. *Zen Buddhism: Selected Writings*. New York: Doubleday.

_____. 1964. *An Introduction to Zen Buddhism*. New York: Grove Press.

_____. 2015. *Selected Works, Vol. 1: Zen*. Berkeley: University of California Press.

_____. 2019. *Zen and Japanese Culture*. Princeton: Princeton University Press.

Teilhard de Chardin, Pierre. 1974. *Christianity and Evolution*. New York: Harvest Books.

Wang, Robin. 2012. *Yinyang: The Way of Heaven and Earth in Chinese Thought and Culture.* Cambridge: Cambridge University Press.

Watson, Burton. 2013. *The Complete Works of Zhuangzi*. New York: Columbia University Press.

Daoist Aspects of Time Perception

in Hakuin's Zen Experiences

QI SONG[1]

Daoism and Buddhism both value meditation, a practice that, after the Song dynasty, became increasingly popular also among Confucians. Although the methods, key features, and goals differ with each group and situation, the time and content that practitioners perceive during the process are highly similar. Meditation is called quiet sitting (*jingzuo* 靜坐) or sitting in oblivion (*zuowang* 坐忘) in Daoism and sitting in absorption (*zuochan/zazen* 坐禪) in Zen Buddhism. It allows practitioners to go beyond time and space as perceived in ordinary reality, transcend life and death, find oneness with the universe, and experience enlightenment (*wu/satori* 悟) by realizing their true original nature (*jianxing/kenshō* 見性) as Dao or Buddha.

In the Southern School of Zen, the Sixth Patriarch Huineng 慧能 (638-713) experienced sudden enlightenment (*dunwu* 頓悟), connecting to the realm of eternity in an instant. Psychologists call this kind of transcendent experience an "altered state of consciousness" (Ludwig 1966; Tart 1969). In the process of practice with the goal of *kenshō* or "seeing one's true nature," that is, experiencing reality as part of a greater, underlying potentiality, in some cases adepts also undergo something called Zen sickness (*chanbing* 禪病).

This comes with two types of symptoms: one consists of psychological obstacles that hinder meditation; the other is a physical ailment caused by over-zealous practice. Key concerns within Zen are accordingly how best to overcome this, how to grasp the present moment while also connecting to the karmic causality of the three periods of life (past, present, and future), and how to adjust the body during practice. Whether or not these issues are resolved has to do with the effectiveness of personal practice and the smooth transmission of instructions.

[1] Translated by Livia Kohn.

Japanese Zen is the direct inheritor of Chinese Chan, transmitted first by Eisai 栄西 (1141-1215) who established the Kenninji 建仁寺 in Kyoto, the first Zen temple in Japan, and vigorously promoted Zen of the Rinzai (Linji 臨濟) school. Later, teachings of other schools, such as Sōtō (Caodong 曹洞) and Ōbaku (Huangbo 黃檗), were also transmitted, leading to tripartite division of Zen in Japan.

In the Edo period (1600-1868), a leading Rinzai patriarch emerged by the name of Hakuin Eikaku 白隱慧鶴 (1686-1769), later mostly called Zen Master Hakuin. His version of Zen attached much importance to historical foundations as formulated during the Song and Ming dynasties. It also showed a great deal of energy and vitality, overcoming the tendency to be overly concerned with words and revitalizing a religion that had withered due to increased prosperity and a close connection to the ruling class (see Luo 2012).

In his spiritual autobiography, *Itsumadegusa* 壁生草 (Wild Ivy; see Waddell 1999), Hakuin records his experiences of *zazen* and *kenshō*, and investigates how his perception of time changed as he underwent the process and experience of sitting meditation. In the following, I make time perception the main focus of inquiry, attaching importance to textual analysis and exploring Daoist elements as found in Hakuin's understanding of Zen, in an effort to analyze the practical significance of his methods for modern society.

Hakuin's Life

Hakuin was born in early 1686 in the village of Hara 原 in Suruga 駿河 county at the foot of Mount Fuji (modern Numazu 沼津, Shizuoka prefecture). His surname at birth was Nagasawa 長澤 and his personal name was Iwajirō 岩次郎. He was born in a business family, but from early childhood showed a unique understanding of religion, supported greatly by his mother who fervently believed in Nichiren Buddhism.

When he was eleven years old, he went with his mother to the local Nichiren center at Shōgenji 昌原寺. Hearing a lecture about hell and its vicious punishments, he became terrified and determined to find a way out of this fate (Waddell 1999, xiv). In due course, at the age of fifteen, he became a monk at Shōinji 松蔭寺 and received the religious name Ekaku. Soon he was transferred to its sister temple Daishōji 大聖寺, also in Numazu, where he made the following vow: "Until this physical body has the power to meet fire and not burn, be immersed in water and not sink, I make endless efforts until I die."[2] From then on, Hakuin never ceased his intense pursuit of enlightenment through deep meditation.

During the seventeen years from 1699 to 1716, he wandered widely throughout the country, visiting numerous prefectures, practicing at thirty-six different temples, and studying with many senior masters. One who exerted great influence on his thinking and practice was Shōju Rōjin 正受老人, originally called Dōkyō Etan 道鏡慧端 (1642-1721). He was the common-law son of Sanada Noboyuki 真田信之, himself the feudal ruler of Shinshu Matsudai 信州松代 (Nagano prefecture). It was under his guidance, at Eiganji 英严寺, that Hakuin had his first *kenshō* experience, but it made him so proud and arrogant that the master refused to issue him *inka* 印可, the official certificate that recognizes the enlightened state and formalizes the right, bestowed by the Buddha and the patriarchs, to teach disciples in his own right. This refusal spurred Hakuin greater efforts, making him a much more powerful practitioner in the long run and a superb teacher when he finally received *inka*.

Another major inspiration was Master Hakuyu 白幽 (d. 1709), a famous hermit who lived in a cave in the Kitashirakawa 北白川 mountains on the east side of Kyoto. Hakuin sought him out when he was suffering from Zen sickness, and the hermit taught him a variety of Daoist methods, from breath control to visualization of *qi*, collectively called "inner observation" (*neiguan/naikan* 内觀). This methodology made a big difference to Hakuin. Not only did it save him from debilitating weakness and add to his spiritual repertoire, but it also deepened his practice and gave him expertise on how best to deal with Zen sickness. Overall, the two masters made him the great master he was and laid the foundation for his unique teaching style and meditation methodology.

In 1718, Hakuin became the abbot of Shōinji and in due course lectured at many temples in the greater Shizuoka area. In addition to visiting numerous Zen institutions, he was also active in writing and painting. He wrote a number of autobiographical records, which were published variously, beginning during his lifetime. His Zen paintings were often humorous and numbered in the thousands: they are characterized by a strong and unique style as well as an inherent explosive power.

In addition, because the language of Zen—being based on Classical Chinese—was hard for ordinary Japanese to understand, Hakuin wrote a primer, entitled *Zazen to san* 坐禅和赞 (Zen Meditation and Dialogue). Its

---

[2] もしこの肉身にして火も焼くこと能わず、水も漂わすことを能わざる底の力を獲ずんば、設い死すとも休まず(Yoshizawa 2016, 28). For more on Hakuin's life, see Katō 1985; Rikugawa 1963; Shaw 1963, 16-19; Waddell 1999, vii-xl.

style is easily accessible to everyone, allowing even ordinary people to master the basics of Zen, and the book has played an important role in the wider promotion of Buddhist practice.

After Hakuin died in 1769, Emperor Go-Sakuramachi (1704-1813) bestowed upon him the posthumous title Zen Master of Spiritual Prowess and Unique Wonder. In 1884, the Meiji Emperor (1852-1912) gave him the honorific position of National Master of the Authentic Lineage (Hakuin 1898, 1:1-2). Soon after, in the Taishō era (1912-1926), special exhibitions about him and his work began to appear, and his popularity has continued unabated to the present day. For example, in 2016, close to the 250th anniversary of his death, the Rinzai temple Ōmotoyama dera 大本山寺 sponsored a series of large-scale rituals in his honor. Around the same time, the Tokyo Museum of Art held an exhibition of his calligraphy and painting, followed by similar events in 2018 at the Kyushu National Museum, the Shizuoka City Art Museum, the Sano Art Museum, and more. They all show just to what degree Hakuin and his teachings are still alive and well in modern Japan.

Kenshō

The word *chan* originally indicated a particular imperial sacrifice to heaven. With the introduction of Buddhism into China and the increased demand for translations of its scriptures, many Buddhist terms were translated into Chinese. The Pali word *dhyāna* means meditation, contemplation, or absorption; it was first rendered phonetically as *channa* 禅那, then abbreviated *chan*. Similarly, *samādhi* refers to a state of mind and body that is fully concentrated, deeply entranced, and one-pointed. It was first transliterated as *sanmodi* 三摩地, then rendered *ding* 定, literally "stability," based on its meaning (Fang 1999).

The two combined to form the compound *chanding*. The general term for Zen practice, it integrates the two dimensions: deep concentration is the desired state of mind, and meditation is the prime method to get there. However, Zen also acknowledges certain peak experiences that occur in a state of deep mental stability. Attained through prolonged and intensive sitting meditation, they are especially *kenshō*, which marks a major breakthrough toward oneness with Buddha-nature, and *satori*, the falling-away of body and mind in complete enlightenment.

For Hakuin, someone who has not experienced *kenshō* and/or *satori* is not a true Zen person. As he says in the *Itsumadegusa*, "Someone who calls himself a man of Zen must first of all experience *kenshō* and *satori*. If someone calls himself a Zen person without such experience, he just presents an imitation in outside appearance" (Yoshizawa 1999, 128-29).[3] In his single-minded pursuit of these altered states of consciousness, Hakuin started to explore different modes of practice when he was young, then continued his work with different masters.

In 1707, when he was twenty-three years old and studying with Unanshi Yomo 云脚四方, he once passed through Banshu 播州 (Hyōgo prefecture) and stayed at a small mountain temple. Seeing the running water of a mountain stream nearby, he was deeply moved and wrote the following poem:

> From the mountain flows a stream,
> On and on, without ever stopping.
> If Zen mind were like this,
> How could *kenshō* be delayed?[4]

This shows that Hakuin saw the persistent, ever-present Zen mind as a necessary precondition for any kind of awakening experience. This Zen mind needs to be ceaseless, without stopping, which means practice must continue at all times, that *zazen* must become part of one's very being and never be interrupted.

Later in the same year, Hakuin returned to Shizuoka and practiced at Shōinji. In November, Mount Fuji erupted—the third major eruption recorded in Japanese history and the first since the Heian period—leading to strong tremors in the temple. All the monks rushed off to safety, but Hakuin remained in the Zendō, stable and firm in his continuous meditation (Takihama 2013, 162). This anecdote illustrates his fearless spirit as well as his willingness to risk life and limb for the sake of Zen practice—based on both, his inherent character and his deep belief in Buddhism.

In the following year, at Eiganji in Echigo (Niigata prefecture), he undertook seven days of continuous meditation and had a major *kenshō* experience, recorded in many of his major works. In his *Orategama* 遠羅天釜 (The Embossed Tea Kettle; dat. 1748), he describes just how depressed and desperate he was before the event:

> In the spring of my twenty-fourth year, I was practicing hard in the Eiganji in Echigo. Day and night I would not rest: I forgot all about food and sleep. Then suddenly, I felt as if I was enclosed by a huge layer of ice that extended

---

[3] 夫れ禅家者流と称せん者は、最初見性道すべし. 若し見性せずして禅人と称せば、紛れも無き大似勢者に非ずや.

[4] 山下に流水有り、滾々として止む時無し. 禅心若し是の如くならば、見性豈に其れ遅からんや.

> into infinity. Although my heart and mind, inside and out, was clean and pure, I found myself unable to move either forward or back. Completely stunned and utterly dazed, I had no words in my heart. I could hear the master chanting at the communal table, but the sound seemed to be coming from a great distance, as if carried about by wafts of air.[5] (Yoshizawa 2001, 9:427)

After experiencing this state of pervasive daze, he realized his first experience of *kenshō*:

> After a few days, suddenly one night, when I was sitting *zazen*, I heard a bell from a distance. After hearing it, I felt as if a plate of ice was smashed to pieces or a jade pavilion was toppling over. When I came to my senses, I realized that I was with the monk Yantou and had passed through all three dimensions of time without any harm.
>
> All doubts of the past disappeared like ice melting into water. I could not help myself but had shout out loud: "This is amazing! This is so amazing!" I realized then that one cannot pursue enlightenment without breaking away from all life and death.
>
> This truth even the 1700 *koans* transmitted from the masters are not enough to decipher. From then on, arrogance rose in me like a mountain as pride surged through my being like a riptide. I felt that for two or three hundred years, no one had been able to achieve such a joyful enlightenment as me.[6] (2001, 9:427-28)

The monk he mentions in this record is Yantou Quanhuo 嚴頭全豁 (828-887), an eminent monk of the Tang dynasty who was killed by marauding bandits at his temple and in due course became the subject of various *koans*. In an earlier period of his life, when Hakuin first heard about his fate, he was deeply disturbed that even an exceedingly moral life and persistent Buddhist practice could not protect one from such a violent death and fell into deep doubts about Buddhism.

---

[5] 二十四歳ノ春、越ノ英巖ノ僧舎ニ在ツテ苦吟ス．晝夜眠ラズ、寝食トモニ忘ル．忽然トシテ大疑現前シテ、萬里一條ノ層氷裏ニ凍殺セラルノガ如シ．胸裡分外ニ清潔ニシテ、進ムコ`ト得ズ、退クコト得ズ、痴痴呆呆、只ダ無ノ字有ルノミ．講莚ニ陪シテ師ノ評唱ヲ聞クト然モ、数十歩ノ外ニシテ堂上の議論ヲ聞クが如ク、或イハ空中ニ在ツテ行クが如シ．

[6] 此ノ如キ者数日、乍チ一夜、鐘聲ヲ聴イテ発転ス．氷盤ヲ擲碎スルガ如ク、玉楼ヲ推倒スルニ似タリ．忽然トシテ蘇息シ来タレバ、自身直に是レ岩頭和尚．三世ヲ貫通シテ毫毛ヲ損セズ．従前ノ疑惑、底ヲ盡シテ氷消ス．高聲ニ叫ンデ曰ク、也大奇也大奇、生死ノ出ヅ可キ無ク、菩提ノ求ムル可キ無シ．伝燈千七百箇ノ葛藤、一捏ヲ消スルニ足ラズ．此ニ於テ、慢幢、山ノ如クニ聳エ、僑心、潮ノ如クニ湧ク．心ニ密カニ謂ラク、二三百年来、予ガ如ク痛快ニ打發スル底、之レ有ル可カラズ．

This was completely resolved in this *kenshō* experience, when he felt that he was able to pass through all kinds of temporal dimensions—phases of life and death—"without any harm" and became one with Yantou. He fully realized the Buddhist teaching of the three stages of cause and effect and understood that Yantou's violent death may have well been the result of an evil deed committed in a previous life. Based on this, he understood that Yantou's fate was not contrary to the Buddhist dharma and fully recognized the principle of causality at work throughout the three dimensions of time. As a result, he was elated and could not restrain his excitement, fully realizing that one cannot be enlightened and attached to past and future at the same time.

Hakuin also records this *kenshō* experience in several other works, each time emphasizing its abrupt occurrence and sudden nature. For example, in the *Sekisho unjō* 隻手音声 (The Sound of One Hand Clapping; dat. 1753), he says,

> I became a monk at the age of fifteen, and between the ages of twenty-two and twenty-three, I developed a great urgency for realization which consumed meday and night. Then, in the spring of my twenty-fourth year, I was at Eiganji in Echigo when I heard the bell sound in the middle of the night, and suddenly I hit the big one.[7] (Yoshizawa 2001, 12:43)

His later work, the *Itsumadegusa* (dat. 1766), has another account.

> Disappointed and bereft of strength, I went to hide myself away in a shrine room dedicated to the lords of the province, vowing to fast and concentrate single-mindedly on my practice for a period of seven days. No one in the temple knew where I was or what I was doing, nor even the monks I had come with. Unable to find me, they assumed I had left secretly for home.
>
> At around midnight on the seventh and final night of my practice, the boom of a bell from a distant temple reached my ears. Suddenly, my body and mind dropped completely away. I rose clear even of the finest dust of the world. Unbearably joyous, I shouted out loud: "Old Yantou is alive and well!"
>
> My yells brought my companions running from the monks' quarters. We joined hands, and they shared with me the intense joy of the moment. After that, however, I became extremely proud and arrogant in my heart. I regard-

[7] 老夫初め十五にして出家、二十二三の間、大憤志を發して、昼夜に精彩を著け、二十四歳の春、越の英厳練若に於いて夜半鐘聲を聞いて、忽然として打發す.

> ed the people I encountered as mere clods of dirt.[8] (Yoshizawa 1999, 172-73; Waddell 1999, 26)

Hakuin attaches great importance to this first experience, which forms an indispensable part of his autobiography. Although there are subtle variations in the different accounts, written a few years apart, the overall description is the same.

Time Perception

The experience was out of this world, instant and sudden, overwhelming and beyond all expectations. Although his epiphany only took a tiny instant in terms of time, it made him feel that he had passed through past, present, and future. It also led to a lasting new way of looking at the world and resulted in long periods of reflection, stimulating modes of thinking that were extremely rich and powerful.

In describing the momentous event, moreover, Hakuin uses the analogies of a plate of ice being smashed to pieces and a jade pavilion toppling over. He also says that doubts he held previously were melting away completely like ice as he experienced great enlightenment, his body and mind were falling off, the ten directions became vast and void, and not even an inch of the planet earth was there any longer. However, the result of this experience did not make him a better person or allow him to enter a higher level of practice. Instead, more obstacles arose and he became proud and arrogant. As he says, "I regarded the people as clods." Even so, judging from the nature of his experience, there is no doubt that this overarching attitude was the result of a genuine *kenshō*. He further notes in the *Sekisho unjō*:

> At the time, there was no fear or anxiety of any kind. After a while, I don't know how long, I penetrated to see the true essence of my own original nature and universal truth arose before my eyes like the bright radiance of wisdom. I have not seen or heard anything like it again for the past thirty years nor felt the amazing and deep joy that arises without being sought. It may be described as being enlightened through *kenshō* in an instant or

[8] 力を落して殿の玉屋の内に隠れて、誓って七日断食接心す．寺中、此の趣向を知るもの無し．同行議してひそかに帰国と為す．満ずる夜半、遥かに鐘聲を聞いて、身心脱落、纖塵を絶す．歓喜に堪ず高聲に叫ぶ、岩頭老人猶お好在なりと．同行、寮中に在つて聞き付け、走り来たり手を把つて互いに相悦ぶ．此より慢心大いに指し起こつて、一切の人を見ること土塊の如し．

.

> understood as a glimpse of rebirth in the Pure Land. But really: there is no Pure Land outside the mind; there is no Buddha beyond of the self.[9] (Yoshizawa 1999, 12:37-38)

Having undergone his first major *kenshō* experience, Hakuin came to the firm conviction that all paradise or enlightenment rested completely within the person, and specifically in the heart and mind. In terms of his time perception, first he underwent a prolonged period of *zazen*, where he patiently and repeatedly worked on *koans* and made a great effort to remain awake and alert. Then the overwhelming insight, like a peak experience as described by psychologist Abraham Maslow (1964) or a mystical experience as reported in Christian and other literature (Underhill 1911; Bucke 1901), occurred in a single moment. He described it as coming about abruptly, a clear case of sudden enlightenment. An epiphany that was initiated by the startling sound of the temple bell, it overwhelmed him completely and turned his world upside-down.

Then, however, a point came when he realized that he had a *kenshō* experience, followed by a phase when his feelings and sense of self were completely blurred and there was no awareness of time at all, what other mystics have described as a state of timelessness (James 1936). He went beyond all life and death and perceived the presence of the monk Yantou who had been dead for many centuries, thus overcoming time. He penetrated the causal relationship between the different dimensions of time in his mind and thereby understood the original nature of the Buddha's enlightenment.

Making this new perspective his own, he further sublimated the time he realized in this state of cosmic consciousness from actual time to a more abstract, philosophical level. Since his time perception was both mystical and abstract, yet also linked to deep emotions and altered states of consciousness, it is difficult to grasp.

---

[9] 此時, 恐怖を生せず, 間もなくはげみ進み待れば, いつしか自性本有の有様を立處に見徹し, 真如実相の慧日は目のあたり（に）現前して, 三十年来未だ嘗て見ず, 未だ嘗て聞かざる底の大歓喜は求めざるに煥發せん. 是を見性得悟の一刹那とも名づけ, 是を往生浄土の一大事とも相伝する事にて, 自心の外に浄土なく, 自性の外に仏なし. 一念不生, 前後際断の当位を往と云ひ, 実相の真理現前の当位を生といふ.

To get a better understanding, let us look back at the time before Hakuin had his *kenshō* experience. He had just arrived at the temple to attend a series of lectures by the famous Ōbaku master Egoku Dōmyō 慧極道明 (1632-1721. However, during discussions with him, Hakuin realized that "he was not the enlightened man he was made out to be" (Waddell 1999, 26). As a result, he was disappointed, depressed, dejected, and confused, a great contrast to the powerful focused mind of meditation he forced himself into. This conflicting mental mode, then, set the stage for the overwhelming connection to original nature.

Zen Sickness

Soon after, Hakuin left the area and went wandering again, to encounter his guiding teacher, Shōju Rōjin, a disciple of Shidō Bunan 至道無難 (1603-1676), who in turn followed Gudō Tōshoku 愚堂東寔 (1577-1661). The latter was a leading abbot of the school's headquarters, Myōshinji 妙心寺 in Kyoto, who seriously dedicated himself to rectifying and rejuvenating Zen practice at the time (Luo 2012). Hakuin admired him greatly and endeavored to follow in his footsteps as represented by his dharma heir.

When the two first worked together, the master assigned Hakuin the famous koan known as "Joshu's Mu" ("Does this dog have Buddha-nature?") to break his pervasive pride and arrogance. Hakuin appreciated this, made good progress, and continued to follow him as his teacher. Eventually Shōju Rōjin was able to bestow *inka* on Hakuin, acknowledging him as a worthy successor of the orthodox lineage of great Rinzai masters. His methods of pure Zen, moreover, inspired Hakuin to develop his own system of *zazen*, emphasizing particularly post-*satori* practice for the sake of greater and more permanent states of enlightenment.

Hakuin was twenty-four when he began to work with Shōju Rōjin. After two years of intensive Zen practice and powerful self-denial, however, he underwent a yet different type of meditative experience, suffering massively from Zen sickness for several years. He records its onset in his *Yasen kanna* 夜船閑話 (Idle Talk on a Night Boat; dat. 1755).

> After that night, when reflecting on my daily life, those two conditions of life, activity and non-activity, had become entirely out of harmony. The two inclinations in me toward finiteness and infinity had become indistinct in my mind. I could not make up my mind to do or not to do anything. So the thought occurred to me that I would like to clothe myself in a lustrous glow and throw off my present life and depart from this world.

> Finding myself in such a state of mind, I set my teeth, fixed my eyes, and determined to forego sleep and food [in favor of incessant practice].
>
> But before I had spent many months in that strenuous way, my heart began to make me dizzy, my lungs became dry, and my limbs felt as cold as if they were immersed in ice and snow.
>
> My ears were filled with ringing like rushing waters of a swift river in a deep canyon. My inner organs felt weak, and my whole body trembled with apprehensions and fears. My spirit was distressed and weary and, whether sleeping or waking, I used to see all sorts of imaginary things, brought to me through my six senses.
>
> Both sides of my body were continually bathed in sweat and my eyes were perpetually filled with tears. I knew then that even if I resorted to famous teachers in every part of the country and searched for great physicians in all parts of the world, none of the hundred medicines would be of any avail. (Yoshizawa 1999, 4:99-100; Shaw 1963, 33-34)

This shows that Hakuin, even after finding the right teacher, still had to face major obstacles and resolve the tension between quiet sitting and outside activities, the different dimensions of practice represented by stillness and movement. To once more experience *kenshō*, he gritted his teeth and pushed himself hard—in many ways too hard, leading to state of body and mind that made practice all but impossible.

The various physical symptoms included dizziness in the eyes, ringing in the ears, upset in the internal organs, coldness in the limbs, as well as sleeplessness, apathy, and hallucinations. In many ways, it was like a nervous breakdown, but there also seem to have been aspects of organ failure and serious physical deficiencies—described by Handa Yōichi as manifestations of neurasthenia and tuberculosis (2012, 50; see also Sharf 2015). In terms of the mind, his state was much like before he had his first *kenshō* experience, and again he had to find a new, powerful master to pull him out of this state.

Time perception during periods of sickness and great stress tends to be the radical opposite of the timelessness of mystical experience. Great emotional pressure, as during a nervous breakdown and when faced with serious disease, is a most radical distortion of time that also impacts concentration and skill (Droit-Volet 2014, 486).

Psychologists in this context acknowledge sleep deprivation and the uncertainty of disease—both part of Hakuin's situation—as major time-inhibiting factors, on par with are being held captive, imprisoned, or hostage as well as feeling rejected, depressed, or anxious. In addition, any form of extended waiting and sensory deprivation—in darkness or isolation—can lead to a feeling of extremely slow-moving time or even chronostasis, the sense that time has stopped (Hammond 2012, 26-43). The same also holds true for various psychiatric disorders, such as ADHD, schizophrenia, autism, and even dyslexia, which all tend to slow down time perception (Noreika et al. 2014, 531).

Daoist Aspects

When Hakuin fell into the depth of Zen sickness, no medical treatments showed any effect. Eventually he decided to seek help from Master Hakuyu, a hermit living in a cave in the eastern mountains of Kyoto who was widely praised as a spirit immortal. In his *Yasen kanna,* Hakuin describes his first impression of the master who looked nothing like a Buddhist monk:

> I could see Hakuyu sitting upright with his eyes fixed straight in front of him. His luxuriant hair reached down to his knees. His face was ruddy and beautiful as the fruit of the jujube tree. He was wearing a large cloth as an apron and was seated on a soft straw mat. (Shaw 1963, 35)

His cavern dwelling, moreover, was about six feet square and held few furnishings or food, but there were three books on a small desk: the Confucian work *Zhongyong* 中庸 (The Doctrine of the Mean), the Daoist classic *Laozi* 老子 (*Daode jing* 道德經), and the Buddhist *Jingang panruo jing* 金剛般若經 (Diamond Sutra of Perfect Wisdom) (1963, 35). These texts are representative of the three great religions of East Asia. They also provided the foundation for Hakuyu's teaching, which essentially consisted of various Daoist methods, from breath control through *qi*-guiding to inner observation.

First, the master took Hakuin's hand to feel the level of his pulses and determined that his inner organs—the core centers of Chinese medicine and Daoist cultivation—were out of harmony because he engaged in too extreme asceticism. The basic Daoist teaching, which comes across here, is that the body is the root of all spiritual development and must not be neglected. On the contrary it should be cultivated or, as Hakuyu puts it, "The man of character always looks after the needs of the body in a reasonable way" (Shaw 1963, 38).

To cure Hakuin's sickness, which he diagnosed as an extreme upward movement of heart-fire (1963, 40), Hakuyu duly taught him basic Daoist principles. He started with yin and yang and their continuous and ever-flowing interchange, moved on to the five organs and their qualities, and eventally reached the rhythmic timing of all life as manifest most obviously in respiration. "The breathing which protects the body and the blood causes them to actively rise and fall in regular motion about fifty times a day," the result of a total of 13,500 in- and out-breaths (1963, 36-37).

Working closely with this natural pattern, the heart beats steadily, internal fire and water move in the right direction, and the organs can do their work properly. As he began to work with this, Hakuin moved away from the timelessness of *kenshō* and the time dilation of Zen sickness and tuned into the natural rhythms of all existence, coming to perceive time as a valuable substratum of organic life.

The key method he used to was basic Daoist breath control as advocated already in the *Zhuangzi* 莊子 (Writings of Master Zhuang), which speaks of breathing all the way to the heels (ch. 6). It was later formalized in healing exercises (*daoyin* 導引), associated with the long-lived Pengzu 彭祖 and often structured with the help of the hexagrams of the *Yijing* 易經 (Book of Changes)—all referenced by Hakuyu (Shaw 1963, 38, 40, 42). In addition, the master also expounded on the teachings of internal alchemy, working closely with the timing of the natural world in conjunction with the rhythms of the body. He spoke of fire and water, dragon and tiger, as well as the active increase or decrease of internal energy movement as controlled by different kinds of fire (1963, 40).

The *Yasen kanna* further records the specific method of meditation Hakuyu used to guide Hakuin toward a profound and efficient way of practice that would not harm the body. It says,

> If the meditator has his four elements out of harmony and feels his body and spirit wearied with labor, he must rouse himself and develop the following visualization to realize complete harmony.
>
> He should see himself in an instant placing a lump of deliciously scented, pure and clear cream—about the size of a duck's egg—onto his head, then feel its subtle and wondrous energy pervade and moisturize his entire being.
>
> Sense how it flows all the way through the head and from there sinks down to soak all the way, reaching the shoulders, elbows, chest, diaphragm, lungs, liver, stomach, and more, until it reaches the bottom of the spine and the buttocks, always and everywhere dissolving all congestion and blockages.
>
> At this time, any blockages and obstructions, any feelings of hardness and pain, all the five kinds of accumulation and six kinds of congestion in the chest will follow the mental visualization downward, gently and softly like water running down a hill, making a pervasive gurgling sound. As this pervades and flows through the entire body, both legs will start to feel warm

> and moist. Eventually it reaches the feet, where it stops.[10] (Yoshizawa 2000, 4:141; Shaw 1963, 43-44)

The four elements mentioned here are those of ancient India: earth, water, fire, and air. They need to be aligned properly and activated in the right temporal rhythm so that, when sitting in meditation and working with the visualization one can "lower the heart-fire to collect in the elixir field (*dantian* 丹田) and flow to the soles of the feet." In other words, adepts are to become aware of their various internal energies, especially those connected to the heart—also the seat of spirit and consciousness—then guide them to the lower abdomen and pelvic floor and from there all the way down to the soles of the feet. The practice is both soft and crisp, gentle and slow, relaxed and focused.

Hakuin applied this successfully to cure his Zen sickness and its various physical symptoms and, in later years, created a form of Zen that actively incorporated Daoist elements. As he himself expresses it in the preface to the *Yasen kanna*:

> This ocean of *qi* [*qihai* 氣海] and elixir field of mine—my hips, my legs, my feet, my soles—altogether are my original face, and this original face also contains certain kinds of sensory features such as nose and nostrils.
>
> This ocean of *qi* and elixir field of mine in totality form the true home of my original destiny, and from this true home certain kinds of information are being transmitted.
>
> This ocean of *qi* and elixir field of mine comprehensively are the Pure Land of my mind-only state, and this Pure Land is strong and solemn.
>
> This ocean of *qi* and elixir field of mine together are the Buddha Amitabha within my body, and this Buddha teaches the dharma.
>
> Always envision yourself returning to your original place, always return to your original place.[11] (Yoshizawa 2000, 4:86; Shaw 1963, 29)

---

[10] 行者、定中四大調和せず、身心ともに労疲する事を覚せば、心を起して応さに此想を成すべし. 譬へば色香清浄の輭蘇、鴨卵の大ひさの如くなる者、頂上い頓有せんに、其気味微妙にして、遍く頭顱の間おうるおし、浸々として潤下し来て、両肩及び雙臂、両乳、胸隔の間、肺肝、腸胃、脊梁、臀骨、次第に沾注し将ち去る. 此時に当て胸中の五積六聚疝癖塊痛、心に随て降下する事、水の下につくが如く、歴々として聲あり. 遍身を周流し、雙脚を温潤し、足心に至て即ち止む.

[11] 我此の気海丹田、腰脚足心、総に是我が本来の面目、面目何の鼻孔かある. 我が此の気海丹田、総に是我が本分の家郷、家郷何の消息かある. 我が此の気海丹田、総に此れ我が唯心の浄土、浄土なんの荘厳かある. 我が此の気海丹田、総に此れ我が己身の弥陀、弥陀何の法をか説くと、打帰へし打帰へし常に斯くの如く妄想すべ.

This shows that Hakuin saw his body as symbolized by central Daoist features such as the ocean of *qi* and the elixir field—major energy centers in the abdomen. He further linked these to notions of original face, true home, Pure Land, and Buddha. In this he integrates Daoist and Buddhist teachings in a powerful, harmonious way and greatly enhances the importance of vital energy (*qi*) and the physical body in Zen practice, complete with an awareness of time as it flows through the body and connects the individual to nature.

Doing so, he applies his own ingenious interpretation of both traditions to establish the connection between concepts and visions radically different in their original context. His methodology, moreover, is based on two concepts: first, that in essence all is one and there are no opposite meanings; second, that all concepts and terms are abstract in nature, leaving room for interpretation and unique apperception.

Integrating the Teachings

However many Daoist terms and practices Hakuin introduced into his system, the fundamentally Buddhist nature of his teachings never changed. As he explains in the *Itsumadegusa*, "I think that my way of doing things may look quite similar to the Daoist ways and generally appear not so much like what Buddhists prefer. Well, this is my Zen" (Yoshizawa 1999, 3:301). In this he confirms that appearances may be deceptive and, while he introduces quite a few Daoist practices into his system, they do not affect his status as a Buddhist teacher. In other words, practicing sitting meditation with a Daoist character has no effect on the essence of its Buddhist nature and ultimate goal, but on the whole creates a particular form of Buddhism that represents his unique, new take on Zen.

How Hakuin sees relationship between Buddhism and his Zen, moreover, is spelled out in the *Orategama*. When the two are criticized as conflicting, he says,

> This is nothing but a way of abandoning *zazen* and slandering quiet contemplation. In general, among all the sages and wise men, anybody possessed of wisdom both in history and today, there has not been a single one who has not relied on contemplation and concentration (*chanding*) to complete the Buddhist path. The three stages of morality (*sīla*), concentration (*samādhi*), and wisdom (*prajñā*) constitute the great network that has held the Buddhist path together since antiquity. Who would dare to despise or ignore

them?[12] (Yoshizawa 2001, 9:250)

As this demonstrates, Hakuin believed that the three disciplines of morality, concentration, and wisdom are central elements of practice that all Buddhist sages and wise men must undertake in the process of attaining enlightenment. They form "the great network that holds the Buddhist path together." Meditation and focused concentration lead to the ultimate goal and, in its specific form of Zen, to "a state of perfect vision that goes beyond all patriarchs and transcends all traditions." In this statement in the *Orategama* he adapts a phrase from the introduction of the Song-dynasty koan collection *Biyanlu* 碧巖錄 (Blue Cliff Record; trl. Shaw 1961; Cleary 1977), compiled by Master Xuedou Chongxian 雪竇重顯 (980-1052).

That is to say, one must rely entirely on oneself and one's own innate qualities, getting away from the buddhas and masters of the past as well as overcoming the power of previous models and examples. Ultimately one must find truth in one's own Buddha-nature while yet always continuing to evolve with the times. Going beyond the patriarchs and transcending traditions is the ultimate secret of freedom of Zen masters, that is, "the supreme great meditation" (Yoshizawa 2001, 9:250).

Hakuin does not propose the pursuit of any firm and unchangeable inherent form of practice but favors an individual and personal evolution that changes over time and matches the natural rhythms both as given by the celestial bodies and manifest within one's own self. The fact that one is not limited by any kind of form or solidity—physical, spatial, temporal—is the key to his way of Buddhist realization.

Conclusion

Taking all this together, Hakuin worked with several different modes of time perception in his various Zen experiences. First of all, during *kensho* time for him collapsed into a single instant and vanished completely, opening him to the underlying cosmic reality of timelessness. Then again, when suffering from Zen sickness, time in his experience was dilated and

---

12 蓋シ斯ク云ヘバトテ坐禅ヲ嫌イ, 静慮ヲ謗ルニシ非ズ. 大凡ソ一切ノ聖賢, 古今ノ智者, 禅定ニ依ラズシテ仏道ヲ成就スル底, 半箇モ亦タ無シ. 夫レ戒定慧ノ三要ハ, 仏道万古ノ大綱ナリ. 誰カ敢テ軽忽セン. 然ルニ向キニ謂ユル禅門ノ諸聖ノ如キハ, 超宗越格, 真正無上ノ大禅定. 擬スル則ハ電轉ジ星飛ブ.

extended into long, drawn-out, and painful segments. Those two represent extreme poles that could also be described as heaven and hell, purity and defilement, or oneness and division.

After learning Daoist techniques from Hakuyu, Hakuin switched to a much gentler and more fluid way of practice, leaving these extremes behind and focusing more on the world of stars, nature, and humanity. Working closely with the natural rhythms, he emphasized their manifestation within the body through breathing and energy guiding, notably in the ocean of *qi* and the elixir field. He adapted his body's patterns to natural time structures as manifest both within and without.

His mind, too, matched this, and he engaged in repeated mental activities as consciousness flowed all through the different parts of the body: head, shoulders, arms, chest, inner organs, hips, legs, and feet. As he incorporated the Daoist way of floating energy through the body in close alignment with natural breathing patterns, he worked increasingly with the ongoing repetition of physical and mental processes, establishing a smooth and harmonious movement of the mind—quite like the mountain stream he so admired in his early years.

This integrated cycle of the mind, moreover, could be objectively measured in time, resulting in a way of working with it that was both productive and manifest. All this allowed Hakuin to activate a different way of perceiving time: rather than overcoming it completely in timelessness or suffering from its vagaries in extreme dilation, he went along with it and used it as a tool to attain internal harmony and mental coherence.

Eventually, from envisioning energy flowing through the body in rhythmic cycles, he moved on to a new level of seeing, finding the true root of enlightenment deep within himself. Doing so, he went beyond time as a historical marker and ancestral agent, attaining the ability to do away with the patriarchs and traditions—a feature strongly emphasized already by Rinzai himself.

Another effect of working with time as natural rhythms was that Hakuin could develop a methodology of Zen practice that was accessible to everyone and easy to teach. This is evident particularly in the *Orategama* and *Yasen kanna*, both works that were addressed less to recluses than to lay followers and aimed to help people in all walks of life.

For example, the *Orategama* contains a letter Hakuin wrote to the feudal lord Nabeshima Naohisa 鍋島直恒 (1701-1749), who had grown tired of the political infighting of his time and wished to recover both peace of mind and physical health. He therefore asked Hakuin for advice on recuperation techniques. To help him, Hakuin in his letter records his own practice experiences plus instructions on the essentials of Zen meditation.

Similarly, the preface of the *Yasen kanna* contains a letter of the Kyoto publisher Ogawa Genhei 小川源兵 on the publication of the book. It describes its contents in terms of guiding *qi* and nourishing vital essence with the goal to enrich people's defensive and protective energies and provide detailed instructions on extending life, addressing key concerns of lay people and making Zen practice accessible to all.

Bibliography

Bucke, R. M. 1901. *Cosmic Consciousness: A Study of the Evolution of the Human Mind*. Philadelphia: Innis & Sons.

Cleary, Thomas. 1977. *The Blue Cliff Record*. 3 vols. Boulder: Shambhala.

Droit-Volet, Sylvie. 2014. "What Emotions Tell Us about Time." In *Subjective Time: The Philosophy, Psychology, and Neuroscience of Temporality*, edited by Valtteri Arstila and Dan Lloyd, 477-506. Cambridge, Mass.: MIT Press.

Hammond, Claudia. 2012. *Time Warped: Unlocking the Mysteries of Time Perception*. New York: Harper.

Handa Yōichi 半田栄一. 2012. "Hakuin no Zen shisō to Nanso no hō" 白隠の禅思想と軟蘇の法. *Hikaku shisō kenkyū* 比較思想研究 38:48-56.

Hakuin Ekaku 白隠慧鶴. 1898. *Chokuitsu shōshū kokushi Hakuin oshō zenji* 敕谥正宗国师白隐和尚全集. Tokyo: Kōyūkan.

Fang Litian 方立天. 1999. *Zhongguo wenhua yanjiu* 中国文化研究. Shanghai: Fudan daxue chubanshe.

Katō Shōshun 加藤正俊. 1985. *Hakuin oshō nenpu* 白隐和尚年谱. Tokyo: Shinbun kaku shuppan.

Ludwig, Arnold M. 1966. "Altered States of Consciousness." *Archives of General Psychiatry* 15.3:225-34.

Luo Shiguang 罗时光. 2012. "You Song-Ming chan dao Baiyin chan: Shi zhi shiba shiji zong Richang sician de bijiao yanjiu" 由宋明禅至白隐禅:10 至 18 世纪中日禅思想的比较研究. Ph. D. Diss., Suzhou University, Suzhou.

Maslow, Abraham H. 1964. *Toward a Psychology of Being*. New York: Van Nostrand Reinhold.

Narushima Ryūhoku 成島柳北. 1985. *Ryūkyō shinshi itsumadegusa* 柳橋新誌壁生草. Tokyo: Benseisha.

Noreika, Valdas, Christine M. Falter, and Till M. Wagner. 2014. "Variability of Duration Perception: From Natural and Induced Alterations to Psychiatric Disorders." In Subjective *Time: The Philosophy, Psychology, and Neuroscience of Tempo-*

rality, edited by Valtteri Arstila and Dan Lloyd, 529-56. Cambridge, Mass.: MIT Press.

Rikugawa Taiun 陸川堆雲. 1963. *Hakuin oshō shōden* 白隠和尚詳伝. Tokyo: Sankikō busshōin.

Sharf, Robert. 2015. "Is Mindfulness Buddhist (and Why It Matters)." *Transcultural Psychiatry* 52.4:470-84.

Shaw, R. D. M. 1961. *The Blue Cliff Records*. London: Michael Joseph.

_____. 1963. *The Embossed Tea Kettle: Orate Gama and Other Works of Hakuin Zenji*. London: George Allen & Unwin.

Takihama Shōjun 瀧瀬尚纯. 2013. *Hakuin shujō honrai butsu nari* 白隠衆生本来仏なり. Tokyo: Heibonsha.

Tart, Charles T., ed. 1969. *Altered States of Consciousness*. Hoboken: John Wiley & Sons.

Underhill, Evelyn. 1911. *Mysticism: A Study in Nature and Development of Spiritual Consciousness*. London: Methuen & Co.

Waddell, Norman. 1999. *Wild Ivy: The Spiritual Autobiography of Zen Master Hakuin*. Boston: Shambhala.

Yoshizawa Katsuhirō 芳沢勝弘. 1999. *Hakuin Zenji hōgo zenshū* 白隠禅師法語全集, Part 1. Kyoto: Zen bunka kenkyūjo.

_____. 2000. *Hakuin Zenji hōgo zenshū* 白隠禅師法語全集, Part 2. Kyoto: Zen bunka kenkyūjo.

_____. 2001. *Hakuin Zenji hōgo zenshū* 白隠禅師法語全集, Part 3. Kyoto: Zen bunka kenkyūjo.

_____. 2016. *Shinpen Hakuin Zenji nenpu* 新编白隐禅师年谱. Kyoto: Zen bunka kenkyūjo.

Synchronicity

A Modern Interpretation of Time in the *Yijing*

JUAN ZHAO[1]

Synchronicity is a key concept developed by the Swiss analytical psychologist Carl Gustav Jung (1875-1961), along others for which he is equally well known such as the personal and collective unconscious, archetypes, individuation, and a particular take on the self. Intertwined with these various concept—that for the most part remain within the framework of Western thought—synchronicity is a mainstay of Jungian psychology. It also relates closely to Eastern thinking, especially as expressed in the *Yijing* 易經 (Book of Changes). This article focuses on the concept of synchronicity and analyzes its relationship with the *Yijing*, then explores it as a modern interpretation of time as understood in ancient Chinese thought, explains it in the context of Western academic terms. Beyond this, as synchronicity involves Jung's reflection on human thought and fundamental life issues, the analysis also opens ways of using the concept of time in the *Yijing* as a principle of universal relevance in the contemporary world.

The Concept Emerges

Jung's early works do not speak of synchronicity. Apparently, the idea first entered his horizon when he came in contact with the physicist Albert Einstein (1879-1955) around 1910. In 1909, Einstein left the patent office in Bern to teach at the Physics Department of the University of Zürich. At that time, his concept of relativity had already gained considerable influence in the scientific community and he was invited widely, including also to Jung's house. Einstein's notion of the relativity and inherent continuity of time and space in due course caught Jung's attention. As he later notes, "It was Einstein who first started me thinking about a possible relativity of time as well as space, and their psychic conditionality. More than thirty years later this stimulus led to my relation with the physicist Wolfgang Pauli and my thesis of psychic synchronicity" (Coward 1996, 480).

---

[1] Translated by Livia Kohn.

The dominant understanding in Western culture since Isaac Newton (1642-1726) was that time was linear and homogeneous, absolute and irreversible. With the arising of the new physics, scientists came to question this, raising issues of time to a key position in both physics and philosophy of the 20th century. Relativity and quantum theory fundamentally upended the concept of time, exerting a powerful influence not only in the natural sciences but also in the humanities and social studies. Time became a key term to understand human existence and culture. Doing away with the basic laws of cause and effect as expressed by René Descartes (1596-1650), thinkers no longer regarded linearity and direct causation as the touchstone of truth. Physicists, philosophers, and cultural anthropologists—who all used to treat time as objective, homogeneous, and irreversible, each from their own perspectives—now pushed its understanding to new limits. They no longer regarded transcending time as the only way to seek the essence of things and the absolute truth of life, but came to observe and experience various—and even contradictory—possibilities in the flow of time.

Jung's concept of synchronicity can be seen as the direct result of this revolution in temporal awareness. However, it also owes a great deal to his exploration of analytical psychology and grew on the basis of practical experience in counseling and theoretical experimentation in psychological analysis. As he says,

> My engagement with the psychology of unconscious processes compelled me many years ago to look around for another explanatory principle, because the causal principle seemed to me insufficient to explain certain strange phenomena of the psychology of the unconscious. That is to say, I first found that there are parallel psychological phenomena that cannot be causally related to each other but must connect in a different context of events. This connection seems to me essential in the fact of the relative simultaneity, hence the term "synchronistic."[2]

In Jung's psychoanalytical work, parallel psychological phenomena or meaningful coincidences often stymied him. Engaging in discussions with the missionary and China scholar Richard Wilhelm (1873-1930) and learning from him about the *Yijing*, he soon came to the understanding that Eastern thinkers did not base their science on strict rationality and regarded coincidence rather than causality as the fundamental principle of how the world worked. As he says himself, "The science of the *Yijing* is

---

[2] Eulogy for Richard Wilhelm at his memorial service, 10 May, 1930 (Wilhelm and Jung 1962, preface).

not based on the causal principle, but on a principle that has not yet been named, because we are not aware of this principle which I have tentatively referred to as the synchronistic principle" (1962, preface).

He continues to point out that synchronism—the term he used before synchronicity—constituted the strength of Eastern thought, while causality formed "the modern prejudice of Westerners." Although he clearly passes judgment in this evaluation, the fact that he modifies the word "prejudice" with the adjective "modern" means that he did not see this as an inherent problem of the West but a shortcoming of the modern age, more precisely, as the conceptual prerequisite for the development of modern science since the Age of Enlightenment. He also, it becomes clear, saw the two as having an equal impact: both causality and synchronism are fundamental principles of thought and form basic factors of experience.

In 1935, in a lecture at the Tower of Dusstock Clinic in London, he first used the word "synchronicity," noting that "Dao can be anything; I use another term to denote this principle, although it is quite basic: I call it synchronicity" (Jung 1935). Still, while the concept now had a name, it remained obscure for quite some time, just dentiong the summary and generalization of all kinds of supernatural phenomena and mysterious experiences. It was, at this time, not yet a core concept in Jungian analytical psychology.

Defining Synchronicity

This changed when Jung connected with the physicist Wolfgang Pauli (1900-1958), who won the Nobel Prize in Physics in 1945 for having developed the Pauli Exclusion Principle. A talented scientist and prolific writer at an early age, praised profusely by Einstein, he was yet a difficult personality, acrimonious and highly critical. His mother had committed suicide, his marriage went wrong, and after a series of setbacks, he suffered a nervous breakdown. Finally, in 1931, he sought out Jung and underwent treatment in analytical psychology. As Jung recorded Pauli's dreams, Pauli soon started to participate in their analysis, criticizing Jung's views from a scientific perspective and increasingly playing an active role in the development of the concept of synchronicity. As Marialuisa Donati says, "In fact, as a consequence of their collaboration, synchronicity was transformed from an empirical concept into a fundamental explanatory-interpretative principle, which together with causality could lead to a more complete worldview" (2004).

The two continued to interact closely until Pauli's death in 1958. As documented in their work, *Atom and Archetype* (2001), they discussed Pauli's dreams as recorded by Jung in their letters and explored the relationship between the inner mind and the outer world in terms of synchronicity, archetypes, and more. Their close cooperation, especially with regard to the notion of synchronicity, is also evident in their book, *Nature and the Interpretation of the Mind* (1952). More specifically, it contains two relevant articles, Jung's "Simultaneous Sensing as the Law of Noncausal Association" and Pauli's "The Influence of the Archetypes on the Formation of Kepler's Natural Science Theory."

Jung's paper in particular is an expansion of his presentation "On Synchronicity," given at the Eranos Roundtable in 1951, the annual meeting of inspiring minds hosted by the Eranos Foundation, based in Ascona, and in that instance focused on the topic "Humanity and Time." Jung notes,

> In this short presentation, I unfortunately can only give a rough outline of the huge issue of simultaneity. For those who want to go deeper into this topic, I would like to suggest that you look to my article on the subject soon to be published in my and Professor W. Pauli's book *Nature and the Interpretation of the Mind.*

Published in the following year in the *Eranos Jahrbuch* (Jung 1952), this article is probably Jung's most concentrated and concise formulation of synchronicity but, being brief and directed at a nonspecialist audience, it inevitably simplifies certain issues. However, it also places the topic into a wider context, especially since at the same meeting Hellmut Wilhelm (1905-1990), Richard Wilhelm's son, presented on the issue of synchronicity from the perspective of the *Yijing* (1952), so that his work and Jung's article form a dialogue and exchange of ideas.

The expanded version in the volume with Wolfgang Pauli, published in the same year, is more intricate. Jung here explores the difficulties and nuances of the problem of synchronicity in as much detail as possible. However, perhaps due to the irrational characteristics of the concept, in many ways he makes it rather complex and creates a certain degree of confusion and misunderstanding. He does not define the concept right in the beginning but, as he says, "I would rather take a different approach and first briefly describe the facts touched by the concept."

After outlining these facts, he classifies them in three key categories that form the backbone of his theoretical discussion as based on empirical phenomena. The three are: associations of events and/or thoughts that cannot be explained by causal associations or have explanations that are invalid, simultaneous events, and coincidences of events and/or thoughts that appear at the same time and can be verified later. In addition, any of these, as he emphasizes, must be relevant to the person and create a "sense of meaning" in the mind.

As already made clear in the title of this expanded paper, he defines the nature of synchronicity as "non-causal association," that is, a law or principle. As a law, it is not only about physics or psychology, but defines the fundamental basis of human experience and knowledge. As he notes in the preface of the book, "My concern for this issue is not only based on scientific research, but on a study of the entire human race" (Jung and Pauli 1952, 2). He points out that, according to modern physics, the laws of nature are only statistical truths and the law of causality is the only certainty, valid in entire range of phenomena. Then he asks, "Does daily experience also show non-causal connections at the macro-physical level?"

Using the parapsychological experiments by J. B. Rhine (1895-1980) as a backdrop, Jung provides powerful illustrations and concludes that under certain spiritual conditions, time and space become relative and can even be overcome. He also proposes a possible mental dynamic to explain how an activated archetype can lead to the emergence of synchronicity. To wit, when the energy of consciousness is relatively low, the energy of the unconscious is relatively strong, and the content of the unconscious will enter the level of consciousness more frequently than usual. There is a certain amount of "absolute knowledge" in the unconscious content, that is, the content of consciousness can go beyond the limits of time and space as shown in parapsychology. If the connection between absolute knowledge and physical events that occur at the same time is noticed and identified, synchronicity will be experienced.

Moving on from there, in Chapter 2 of the book, Jung describes a set of astrological experiments. He concludes that, if the relationship between astrological expectations and the final result is not causal, it must contain a certain meaning. In Chapter 3, he traces the origin of synchronicity both in Chinese and Western thinking but in the final chapter he admits that the entire concept has not been fully proven or reliably confirmed. He still tentatively proposes it on the basis of empirical observations of events that occur outside the body and are nonlocal and notes that the relationship between body and mind may be one of simultaneous feelings. He also elaborates the theoretical status of synchronicity as a fourth explanatory principle—that is, he posits the "principle of synchronicity" in addition to time, space, and causality.

To sum up, Jung's multifaceted discussion of synchronicity includes the following key aspects. Two or more events are not related by cause and effect, but have the same or similar meaning and coincide in time. Related in creative action, they are noncausal yet parallel. Synchronicity, therefore, means the coincidence of two or more thoughts and/or events that may occur at the same time or be confirmed later as having occurred simultaneously. They are independent of place, that is, nonlocal, and there is no causal connection between them. The law of causality cannot explain how they come about, but the coincidence attracts attention because the events have the same or similar meaning. If all this holds true, the connection between the events is one of synchronicity. Four key words summarize it best: time, non-causality, coincidence, and meaning. In addition, the concept also represents an echo of the revolution in the understanding of time in the new physics and a modern reflection of the 20th century.

The *Yijing*

In Jung's lifetime, the *Yijing* had already been translated into English by the missionary and sinologist James Legge (1815-1897; see Girardot 2002) as part of his magnum opus, *The Sacred Books of China* (Legge 1899). However, in Jung's view, this translation did not offer an adequate understanding to Westerners. As far as we can tell, he first learned about it from Toni Anna Wolff (1888-1953), a patient of his from 1910 to late 1911 and later, starting in 1913, his mistress and collaborator. He may well have learned about Eastern philosophy and religion from her father who had a keen interest in China. While this is conjecture, by 1920, as his autobiography shows, Jung had definitely begun to use the *Yijing* in his explorations of the human mind (Rong 2009, 320).

However, his dedicated interest in the *Yijing*, as well as its impact on his vision of synchronicity, is due largely to his cooperation with the sinologist Richard Wilhelm. Wilhelm, in China as a Protestant missionary, had spent over ten years translating this difficult classic into German. He worked in close collaboration with Lao Naixuan 劳乃宣 (1843-1921), a senior Qing scholar who served chief education officer of Jiangning and later became president of Hangzhou Qiuzhi University. Lao in turn recommended that Wilhelm should base his work on the exegesis by the renowned thinker Li Guangdi 李光地 (1642-1718), entitled *Zhouyi zhezhong* 周易折中 (Analytical Edition of the Zhou [Book of] Changes) (see Li 2012).

In order to facilitate a good understanding and easy acceptance of the text by European readers, especially also non-professionals, Wilhelm carefully arranged the German translation. He included not only the main text with its different parts and various early supplements—the so-called Ten Wings—but also important interpretations from Chinese commentaries. In addition, he took great care to explain just how the arrangement of the translation worked in relation to the original structure of the *Yijing*.

This undoubtedly made it easier for Europeans to immerse themselves in the work and over time opened the way for them to enter more closely into the Chinese mind (Zhouyi 2010)

The work was published in Jena in 1924 and in due course became the blueprint for all other Western renditions. Its influence was immediate and powerful, especially since European society at the time—in the wake of World War I—was in the throes of strong feelings that civilization was coming to an end, expressed powerfully by the German historian Oswald Spengler in *The Decline of the West* (1922). The Chinese classic offered one way for intellectuals to envision the reinvention and renewal of European culture with the help of non-Western thought—in a way very similar to the reception of the *Daode jing* (see Hardy 1998).

In Jung's view, Wilhelm unlike Legge really grasped the vitality of the *Yijing* and made it perfectly accessible to Westerners. Aware of his work even before the book was published, in 1923 he invited Wilhelm to give a lecture at the Psychology Club in Zürich. At this occasion, the two had plenty of opportunity to discuss Chinese philosophy and issues of religion in general as well as the *Yijing* in particular. This exchange made Jung realize just how powerful the Chinese philosophical tradition was and how much it had to offer his understanding of the world, especially in the light of the problems imposed by the European subconscious mind (Wei and Rong 1993, 140-152). A friendship resulted that, over the next few years, led to the joint translation of the Qing-dynasty Daoist text *Taiyi jinhua zongzhi* 太乙金華宗旨 (The Secret of the Golden Flower), which was published in 1929 and has also had a profound impact on Western thought (Wilhelm and Jung 1962, preface).[3]

So, what exactly did Richard Wilhelm teach Jung about the *Yijing*? To him, the text was a book of divination and wisdom, a concentrated expression of the "Chinese mind and spirit." This certainly echoed with Jung. As a classic representative of Eastern civilization, complete with a long history and wide-reaching influence, the *Yijing* to him was a cultural archetype, an expression of the unique potentiality of human thought. Jung was increasingly convinced that the accidental coincidences he often encountered in the process of analyzing dreams were not just an issue of statistical probability, but revealed a fundamental issue related to human thinking. The *Yijing* presented him with a way of understanding the world

---

[3] For a more recent English translation, see Cleary 1992; for modern Chinese, see Wei and Rong 1993. On the textual history of the work, see Esposito, 1998; Mori 2002. For its importance in internal alchemy today, see Wang and Bartosh 2019.

in terms completely alien to traditional Western models, a way that history and experience had already proven relevant and highly effective.

Still, however much he accepted the *Yijing* and recognized its value, Jung maintained his culturally determined perspective and was always limited by his existing cognitive framework. That is to say, whether he approved or criticized, he remained within the boundaries of analytical psychology. As he said in a letter to B. V. Raman on the subject of astrology, dated 6 September 1947:

> As a psychologist, I am mainly concerned with the coincidence of certain astrological conditions and personalities. In some difficult treatments, I often have astrological signs to get a completely different way of thinking. . . I admit that this is a very strange fact, which gives us a special care for the structure of the human spirit. (Main 1997, 81)

In Jung's view, divination worked in such a way that, even if "there was no event that triggered synchronicity, at least it would make events fall in line to fulfill its purpose." He also notes that "thinking based on the principle of synchronicity reached its peak in the *Yijing*. It can be said that it is a manifestation of the authentic Chinese way of thinking."

Jung's Understanding

Jung's understanding of the *Yijing* is made clear in several distinct sources. In chronological order, they are:

- —his eulogy for Richard Wilhelm at the memorial service in 1930 (Wilhelm and Jung 1962, 139-51);
- —his letter to B. Baur on 19 January 1934 (Jung 1973);
- — his "Foreword" to the English translation of Wilhelm's *Yijing* translation, written in 1949 (Wilhelm 1950, xxi-xxxix);
- —his letter to Rev. W. P. Witcutt on 24 August 1960 (Jung 1990);
- —his autobiographical memoir, dated to 1963 (Rong 2009)

Among all these, the "Foreword" reflects on the relationship between the text and Jung's vision of synchronicity in the most concentrated and detailed manner. Originally written in German, its English version is not the same. While the second half is roughly identical, the first half in German describes Richard Wilhelm's take on *Yijing* divination while that in English discusses synchronicity. Also, the English version was published around the same time as Jung's essays in the *Eranos Jarhbuch* and his joint book with Wolfgang Pauli, the three coming together to represent his outlook.

A key feature of Jung's "Foreword," then, is that he engages in a dialogue with the *Yijing* as "an ancient book that purports to be animated" (Wilhelm 1950, xxvi). Asking its judgment about its present situation, he receives the hexagram Ding 鼎䷱ (Cauldron, No. 50), which he takes to mean that the text considers itself a culturally determined container of spiritual nourishment. Analyzing the various images and lines of the hexagram, he comes to the conclusion that "the *Yijing* faces its future on the American market calmly" (1950, xxxiii).

He then proceeds to ask about his own behavior and position in this greater context and receives the hexagram Kan 坎䷜ (The Abysmal/Water, No. 29). The oracle gives a warning of deep and abysmal danger and advises patience and restraint, "otherwise you will fall into a pit in the abyss" (1950, xxxv). Jung reads this less as a personal warning to stay away from the text than as a commentary on people's relationship to it, the tendency to jump to conclusions when there are unfathomable depths to be explored.

> In the dreamlike atmosphere of the *Yijing*, one has nothing to rely upon except one's own so fallible subjective judgment. Ii cannot but admit that this line represents very appropriately the feelings with which I wrote the foregoing passages. (1950, xxxv)

Along a different line of inquiry, Jung also expresses surprise that "such a gifted and intelligent people as the Chinese never developed what we call science" (1950, xxii). To him, this just shows that the foundation of modern science in the West, that is, the law of causality, is encumbered by limitations, one of which makes it very difficult for Europeans to understand Eastern perspectives. There are fundamental divergences in the way of thinking, most specifically the contrast between causality and synchronicity—the latter most pertinently expressed in the *Yijing*, the most concentrated embodiment of Chinese thought.

For Westerners, this way of thinking has a subversive significance: it shakes their very cognitive foundation. Jung calls this the Archimedes Point, the transformative hinge, where it becomes clear that the law of causality is only a relative truth and has its limitations. Not only does it not work at all in some areas of the mind, but it is also nothing but statistical probability and therefore must face exceptions. Any laws of nature studied under strictly limited conditions in the laboratory, even if they can be verified repeatedly, yet never fully match in reality—each individual's reality is full of processes that are more or less subject to chance. This chance, this oddness, this unpredictability is what Europeans are trying to avoid or, ideally, eliminate completely. Yet, it constitutes a matter of great concern in the Chinese mind. As he says in his commentary to *The Secret of the Golden Flower*,

> The Chinese did indeed have a science, whose standard work was the *Yijing*, but the principle of this science, like so much else in China, was altogether different from our scientific principle. (Wilhelm and Jung 1962, 142)

This altogether different principle is defined in his concept of synchronicity. Jung believes that in this way of thinking, "events fit in time and space and do not just occur by chance. It contains more meaning, in a nutshell, that is, the objective events are between each other and between them and observation There is a special interdependent relationship between the subjective mental states of the readers" (Jung 2000, 209).

He believes that the sixty-four hexagrams of the *Yijing* represent particular life situations, complete with its specific state of mind. As he says,

> Now, the sixty-four hexagrams of the *Yijing* are the instrument by which the meaning of sixty-four different yet typical situations can be determined. These interpretations are equivalent to causal explanations. Causal connection is statistically necessary and can therefore be subjected to experiment. Inasmuch as in situations are unique and can nor be repeated, experimenting with synchronicity seems to be impossible under ordinary conditions. In the *Yijing*, the only criterion of the validity of synchronicity is the observer's opinion that the text of the hexagram amount to a true rendering of his psychic condition. (Jung 2000, 209-10)

In other words, the seeker shares his state of mind by presenting a question, then interacts with the oracle through milfoil stalks or sacred coins, while also integrating the characteristics of his particular situation. Time and space are relativized and subjective to spiritualization, and the law of causality loses its validity. At this point of synchronous interaction, mind, world, and oracle coincide: the *Yijing* opens the human mind to being inspired by deep meaning and engaging holistically with past, present, and future in an experience of mutual presentation and total integration.

However, for Westerners, the fundamental understanding of life as flowing and potentially synchronous was lost. "This way of thinking disappeared from the history of Western philosophy after Heraclitus [ca. 535-755 BCE], but there were some low echoes from Leibniz [1646-1716]." This is why Jung associates synchronicity with Leibniz's concept of "pre-established harmony" (*harmonia praestabilita*). Jung believed that this kind of thinking "has not completely died out in the history of Western philosophy. It is still alive in the fading astrology metaphysics, and it has been preserved to this day."

Some scholars in the West have questioned this view. Donald Phillip Verene, for example, points out that although Jung's concept of synchronicity is mainly based on the divination system as represented in the *Yijing*, this does not mean that there is no support for it in Western philosophy or that there are no theoretical resources for it as a fundamental principle. He believes that the theories of historical experience and personal self-knowledge as proposed by the Italian philosopher Giambattista Vico (1668-1744) are a form of metaphysics based on the idea of synchronicity. He also argues that his notions of language and historical cycles became the basis of James Joyce's (1882-1941) final work, *Finnegans Wake*. Vico's metaphysical understanding of synchronicity and Joyce's cultural expression to him provide a resource on the principle of synchronicity in modern Western thought that has nothing to do with divination. Verene believes that Jung did not pay attention to this resource (2002).

As Jung reflected and criticized the law of cause and effect, he rejected Western ideological resources as useless and directly turned to the *Yijing* as a foundation stone of all ideas of synchronicity. In this endeavor he favored an intellectual approach and did not get involved in passionate debates. His position is thus quite different from those of early missionary sinologists who attempted to find some dimension of divinity and confirm the existence of God in the text of the *Yijing*. It is also very different from that of professional China scholarship in the modern academic sense, since he used the text toward his own ends and read his own visions into it.

It can even be said that the core of Jung's focus was not the *Yijing* as such or the goal of obtaining reliable knowledge about it. Nor was he primarily interested in figuring out the relationship between the text and Chinese culture and spirit. His main concern was how much proof he could find in the text for his own ideological system, and he studied with keen awareness the kinds of situations it described and the means it used to resolve potential conflicts. To him, the key issue was how the life situations and advice given in the *Yijing* could explain and resolve the problems posed by analytical psychology. It was, in other words, only one of many astrological and divination works, both ancient and modern, Chinese and Western, that Jung engaged with in his work. However, it stands out in that it presents one of the most mature and concentrated interpretations of psychological issues and life-situation problems in world history.

Concepts of Time

How, then, does the *Yijing*, which to Jung represents the "authentic Chinese way of thinking," work with the issue of time? In his early work on the subject, including his Eranos presentation, he describes the phenomenon of simultaneous meaningful occurrences with the term "qualitative time," to be later replaced with synchronicity. He clarifies the distinction between the two in a letter to André Barbault, dated 26 May, 1954.

> [Qualitative time] is a notion I used formerly, but I have replaced it with the idea of synchronicity, which is analogous to sympathy or correspondentia, or to Leibniz's pre-established harmony.
>
> Time in itself consists of nothing. It is only a *modus cogitandi* used to express the flux of things and events, just as space is nothing but a way of describing the existence of a body. When nothing occurs in time and when there is no body in space, there is neither time nor space.
>
> Time is always and exclusively "qualified" by events just as space is by the extension of bodies. But qualitative time is a tautology and means nothing, whereas synchronicity (not synchronism) expresses the parallelism and analogy between events in so far they are noncausal.
>
> In contrast, qualitative time is a hypothesis that attempts to explain the parallelism of events in terms of cause and effect. But since qualitative time is nothing but the flux of things, and is moreover just as much nothing as space, this hypothesis does not establish anything except the tautology: the flux of things and events is the cause of the flux of things. (Jung 1990, 16)

While working with the concept of qualitative time, Jung still tried to explain certain phenomena such as significant coincidences within the overall framework of causality. However, once he moved on to the use of synchronicity, it becomes clear that he abandoned this altogether and placed all attempts at interpretation and methods of analysis within a framework of causality into a completely different category. How this works is shown best in the following diagram (Jung and Pauli 1952):

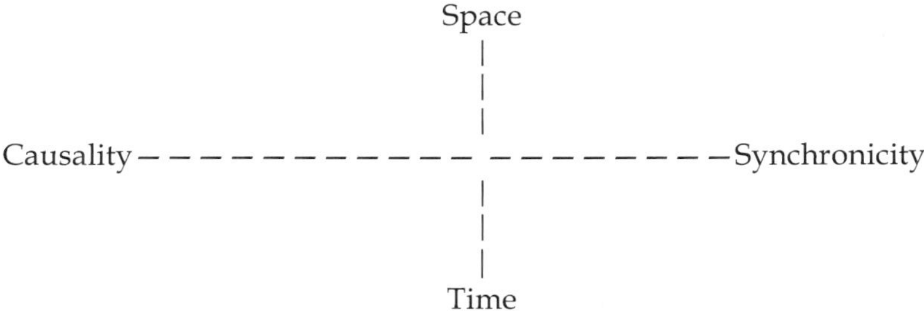

Here causality and synchronicity appear in a relatively equal position, representing two basic laws of how the world is connected in time and space. However, as his discussions with Wolfgang Pauli went on, Jung absorbed some of his suggestions and revised this chart to include notions of the fundamental energy and constancy of spacetime as well as more abstract visions of connectivity and valuation. It then came to look like this:

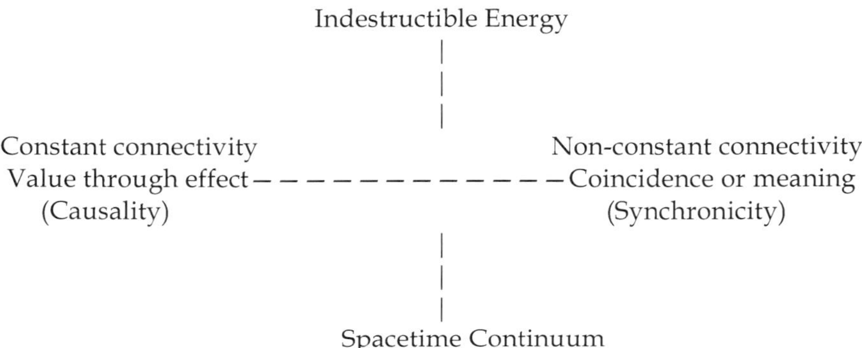

Seen from this angle, if the law of causality is understood as the principle of association between things and events, then time is composed of a series of causes and effects. In other words, things or events occur in the sequence of time, with the cause coming first and the result second and the two forming a temporal chain. Her the spatial vision of sequence—traveling from one place to the next—is the conceptual premise of the law of causality and thus creates a particular perception of time.

In contrast, the concept of synchronicity as a way of thinking manifests outside of time and space. This was further developed by Jung's students and followers. For example, his friend and biographer, Barbara Hannah expanded the theory from the perspective of space and time (1982).

Also, in 1952, Jung engaged in a dialogue with the religious historian Mircea Eliade (1907-1986) who pointed out to him that synchronicity marks a break in time, quite like the experience of the holy in religion: here too time, space, and causality lose their effectiveness (McGuire and Hull 1985). As time is broken, the basis of all causal connections and functions is gone, a phenomenon Roderick Main later called the "rupture of time" (2004).

Jung himself describes this in his memorial for Richard Wilhelm:

> It seems indeed as though time, far from being an abstraction, is a concrete continuum that contains qualities or basic conditions, which manifest themselves simultaneously in various places in a way not to be explained by causal parallelism as, for example, in cases of the coincident appearance of identical thoughts, symbols, or psychic conditions. (Wilhelm and Jung 1962, 148)

The *Yijing* is a great case in point for this vision, mainly because it works with a concept of time that closely matches Jung's synchronicity. It presents a complete system including the seeker, his or her key question, the formal ceremony, the text of the *Yijing*, its interpretation, and the ultimate prediction or advice as a result of the powerful interactive process. While it served as a mainstay of his overall theory, a way of completing his reflections on European modernity, it also can be interpreted from Jung's view: by engaging his concept of synchronicity we may come to see a new and dfferent level of what exactly time means in the ancient Chinese text. Two points stand out: the nature of reality as unrepeatable experience and the hexagrams as snapshots of situations in time plus their relationship to the images found in dreams and the unconscious.

Reality as Experience

The way the *Yijing* looks at reality disfavors causal attitudes and processes. In the ancient Chinese view, each single moment under observation is more of a chance hit than the clearly defined result of concurring causal chain processes. The matter of prime interest is the configuration of thoughts and events formed by coincidental encounters at the moment of observation rather than the hypothetical reasons that apparently account for the situation. While the Western mind carefully sifts, weighs, selects, classifies, and isolates, the Chinese picture of the moment encompasses everything down to the minutest nonsensory detail, because all these different ingredients constitute the observed moment (Jung 2000, 207-08).

Therefore, every instance of casting the *Yijing* is sacred and mysterious to both the seeker and the diviner. Repeated divisions of milfoil or throws of coins may provide different results, but this will not cause people to lose faith in the text's predictions. Plus, if questioned repeatedly on the same issue, the tendency is for hexagram Meng 蒙 ䷃ (Youthful Folly) to appear, which says right in the Judgment: "The young fool seeks me. At the first oracle I inform him. If he asks two or three times, it is importunity. If he importunes, I give him no information" (Wilhelm 1950, 21). In other words, the text refuses to deal with the same question twice, making sure people do not pester it on the same issue.

While the *Yijing* safeguards against conflicting results, its reading might be at odds with those of a different mode of divination, notably that by cracks (*buzhan* 卜占), the most common and best-known method of the Shang dynasty, documented in the so-called oracle bones. Here, diviners would ask a question and, in a formal ritual, drill tiny holes in the carapaces of turtles or shoulder bones of cattle. Heated over a fire, they would produce cracks that, depending on their direction and length, were read as a yes or no response (Keightley 1978). The *Yijing,* in contrast, worked with "stalk divination" (*zhanzhan* 筮占), the counting and dividing of milfoil. This was more common in the Zhou dynasty, as documented in several recently discovered manuscripts, which describe how trigrams, and not just hexagrams, were connected to seasonal patterns and read in terms of good or bad fortune (Cook and Zhao 2017).

On occasion someone would use both major modes of divination and come up with conflicting results. For example, the ancient chronicle *Zuozhuan* 左傳 (Mr. Zuo's Commentary [on the Spring and Autumn Annals]) describes how Duke Xian of the feudal state of Jin, when he wanted to take Princess Li as his wife, divined the outcome in both modes. The cracks gave a negative answer, while the stalks considered the match "auspicious." The duke favored the stalks, but the diviner insisted that turtles lived much longer than grasses and were accordingly holier: they must, therefore, be obeyed (Duke Xi, 4th Year). In other words, when conflict arose, the resolution was not reached by a statistical comparison of the reliability of the two methods, but by considering the nature of the divining medium and its divine standing within the overall culture, a reasoning quite baffling to the scientific mind.

Both the situation of the seeker and the particular presence of the divining medium, therefore, are unique and unrepeatable, manifesting a particular combination of factors and events, thoughts and emotions. They come together at a certain point in time, create a connection, and determine what steps to take or what route to follow. Issues of why and how and where from do not enter the picture. Immediacy and simultaneity triumph over reasoning and analysis.

Hexagrams as Snapshots and Images

Every time someone consults the *Yijing*, the result is a hexagram, won through the consecutive division of pile of forty-nine milfoil stalks or the repeated toss of three coins. The process integrates the various factors of heaven, earth, and humanity into one situation, merging the mind and the external world, the individual and the gods or universe so they form one whole and can communicate jointly. The hexagrams as much as the combined situation are based on specific terms of time and nature, and only thereby can they obtain inherent unity. Time is a major factor in the entire process, and the hexagram shows a great deal of relevant information in this regard, reflecting the past, creating a vivid focus of the present, and pointing out the direction of future developments and the best way to deal with them.

Thus, when one throws the three coins or counts through the forty-nine milfoil stalks, details of chance or coincidence enter into the picture at the very moment of divination and form a part of it—a part that may be insignificant to Westerners yet constitutes the most meaningful aspect in the Chinese mind. For modern Westerners, it would be a banal and almost meaningless statement, at least on the face of it, to say that whatever happens in a given moment is what happens in that instant (Jung 2000, 208).

This instant marks the quality of time in the *Yijing*. It is not linear and homogeneous, has nothing in common with the absolute time of traditional physics. Often, in response to a question, the text will speak of time lost or gained, appropriate or inauspicious. It may advise the seeker to work with time or convey information in accordance with it. It may also recommend to be in sync with time, wait for the right time, and the like. While different people and different events may receive completely different answers, time in all cases has its own unique quality, closely related to the situation as a whole.

This is highly personal time, embedded deeply in the individual and his or her situation, complete different from scientific time that exists objectively outside of either person, thought, or event and does not change according to any particular circumstance. *Yijing* time exists actively in the seeker as a unique assembly of personal, social, natural, and cosmic factors: it integrates the human mind and the entire world into one unified, flowing, and dynamic entity.

People's position is to be placed deeply in this time. Stated differently, time here marks, or even defines, the particular situation in which they find themselves. The question asked is not, "What is time?" but: "What does time signify?" What exactly is the impact that time has in this very moment? What are the temporal markers of the current situation? What behavior or strategy best matches this time?

All these are mysteries revealed through the image of the hexagram, a symbol that contains the will of the gods and reveals the inner workings of heaven and earth. It is a bridge between the visible and the invisible. As Stephen Karcher notes,

> Time: moment, right time, propensity. All these refer to the internal energy or intensity of a situation, a configuration, gesture, stance, or attitude that guides the vital movements of beings. It is the potential, the right time or right season to discern and seize the favorable by regarding the signs of change, signs from heaven that describe the activity of spirit or the gods who notify and instruct us. (2003, 12)

As reality is perceived as the specific development of things at a certain moment, its organization and arrangement depend on the degree to which people grasp this moment: understand it, appreciate its significance, work with it, let it unfold, or hold back on it. The hexagrams of the *Yijing*, marking sixty-four distinct types of moments, therefore, are snapshots that make hidden currents visible. With their help the seeker can understand the subtle tendencies and providential trends of heaven and earth relevant to his or her situation in that very particular moment of time.

To Jung, the system of the hexagrams with their lines, images, judgments, and readings was very much like the analysis of dreams with their images and subtle indicators through creative imagination. As he outlines in a letter to Reverend W. P. Witcutt (24 August 1960), they all connect to a part of the human self or psyche that is not subject to time and space (Jung 1990, 584). Dreams to him are primarily the domain of the unconscious, which in turn divides into the individual and the collective unconscious. The main content of the collective unconscious manifests as archetypes, rooted deeply in human physiological structures and deep cultural conditioning. They appear in numerous archetypal images that people then see in their dreams.

To some extent, the sixty-four hexagrams of the *Yijing* are similar to Jung's archetypal images. In both cases, the validity of any judgment or interpretation, that is, the reliability of defining the key factors of the present and predicting the future, lies in the coincidence between the inner mind and the outer symbols or images. This coincidence makes time relative so that the past flows into the present, the present extends into the future, and the future in its turn conveys information to the present. The irreversible characteristics of past, present, and future in time conceptualized as linear are eliminated completely in this system.

Conclusion

Jung freely admitted that he was no China specialist and had never visited the country. He used the translation of Richard Wilhelm to read and work with the *Yijing*, relating its thought to his concept of synchronicity. For him, the text represented more a revelation of a completely different kind of thinking than an academic research project. He paid close attention to its vision and interpreted it in the light of his understanding of synchronicity as an alternative principle to causality. This reflects his unique sensitivity as a thinker and his intuitive ability to penetrate deeply into the text. Since, like an *Yijing* judgment, his understanding was highly personal, one cannot judge whether it is correct or accurate, but must analyze how Jung himself came to it and what problems he saw. His views then can help us become aware on our own understanding of what issues we have both with working with time in a more general sense and with activating it in our own very specific context.

Over the past century, the cultural and philosophical exchange between China and the West has increased massively, resulting in mutual stimulation, questioning, and learning. The development of Chinese and Western thought has become more and more constructive in relation to each other. Thus, the *Yijing* has become highly meaningful in the context of Western thought and culture, just as its concept of time has turned into an important resource for Jung and other analytical psychologists. The concept of synchronicity is a keystone in the larger revision of temporal awareness in Western academic circles, a cornerstone in the overall development of academic exchange and intellectual dialogue between East and West.

Bibliography

Cleary, Thomas. 1992. *The Secret of the Golden Flower: The Classic Chinese Book of Life*. San Francisco: Harper.

Cook, Constance. A., and Lu Zhao. 2017. *Stalk Divination: A Newly Discovered Alternative to the I Ching*. New York: Oxford University Press.

Coward, Harold. 1996. "Taoism and Jung: Synchronicity and the Self." *Philosophy East and West* 46.4:477-95.

Donati, Marialuisa. 2004. "Beyond Synchronicity: The Worldview of Carl Gustav Jung and Wolfgang Pauli." *Journal of Analytical Psychology* 49:707-28.

Esposito, Monica. 1998. "The Different Versions of *The Secret of the Golden Flower* and Their Relationship with the Longmen School." *Transactions of the International Conference of Eastern Studies* 43:90-109.

Girardot, Norman J. 2002. *The Victorian Translation of China: James Legge's Oriental Pilgrimage*. Berkeley: University of California Press.

Hannah, Barbara. 1982. *C. G. Jung: Sein Leben und Werk*. Fellbach-Oeffingen: Bonz.

Hardy, Julia. 1998. "Influential Western Interpretations of the *Tao-te-ching*." In *Lao-tzu and the Tao-te-ching*, edited by Livia Kohn and Michael LaFargue, 165-88. Albany: State University of New York Press.

Jung, C. G. 1952. "Über Synchronizität." *Eranos-Jahrbuch* 20:271-84.

_____. 1969 [1935]. *Über die Grundlagen der analytischen Psychologie: Die Tavistock Lectures*. Zürich: Rascher & Co.

_____. 1973. *Letters 1, 1906-1950*. Edited by Gerhard Adler with Aniela Jaffé. Translated by R. F. C. Hull. Princeton: Princeton University Press.

_____. 1990. *Letters 2, 1951-1961*. Edited by Gerhard Adler with Aniela Jaffé. Translated by R. F. C. Hull. London: Routledge.

_____. 2000. *The Psychology of Oriental Meditation: From the Book of Changes to Zen*. Translated by Yang Rubin, Beijing: Social Sciences Archives Press.

_____, and Wolfgang Pauli. 1952. *Naturerklärung und Psyche*. Zürich: Rascher & Cie.

_____. 2001. *Atom and Archetype*. Edited by C. A. Meier; translated by David Roscoe. Princeton: Princeton University Press.

Karcher, Stephen. 2004. *Total I Ching: Myths for Change*. New York: Time Warner Books.

Keightley, David N. 1978. *Sources of Shang History: The Oracle-Bone Inscriptions of Bronze Age China*. Berkeley: University of California Press.

Legge, James. 1899. *The Sacred Books of China: The Texts of Confucianism*. Oxford: Clarendon Press.

Li Xuetao 李雪涛. 2012. "Wei Lixian Yijing de yiben de fanyi guocheng ji diben chutan 卫礼贤易经德译本的翻译过程及底本初探. *Shijue hanxue* 世界汉学 9.

Main, Roderick. 1997. *Jung on Synchronicity and the Paranormal*. London: Routledge.

_____. 2004. *The Rupture of Time*. London: Routledge.

McGuire, W., and R. F. C. Hull. 1985. *C. G. Jung Speaking: Interviews and Encounters*. Paris: Editions Buchet Chastel.

Mori, Yuria. 2002. "Identity and Lineage: The *Taiyi jinhua zongzhi* and the Spirit-Writing Cult to Patriarch Lü in Qing China." In *Daoist Identity: History, Lineage, and Ritual*, edited by Livia Kohn and Harold D. Roth, 165-84. Honolulu: University of Hawaii Press.

Rong Ge 荣格 [Carl Jung]. 2009. *Rong Ge zizhuan: Huiyi, meng, sikao* 荣格自传：回忆、梦、思考. Translated by Liu Guobin 刘国彬 and Yang Deyou 杨德友. Shanghai Sanlian chubanshe.

Spengler, Oswald. 1922. *The Decline of the West*. London: George Allen & Unwin.

Verene, Donald Phillip. 2002. "Coincidence, Historical Repetition, and Self-Knowledge: Jung, Vico, and Joyce." *Journal of Analytical Psychology* 47:459-78.

Wang, Liping, and Mark Bartosh. 2019. *Daoist Internal Mastery*. St. Petersburg, Fla.: Three Pines Press.

Wei Lixian 卫礼贤 [Richard Wilhelm] and Rong Ge 荣格 [Carl Jung]. 1993. *Jinhua yangsheng bizhi yu fenxi xinli xue* 金华养生秘旨与分析心理学. Translated by Tongshan 通山. Beijing: Dongfang chubanshe

Wilhelm, Hellmut. 1952. "Der Zeitbegriff im Buch der Wandlungen." *Eranos-Jahrbuch* 20:321-48.

Wilhelm, Richard. 1924. *I Ging: Das Buch der Wandlungen*. Jena: Diederichs.

_____. 1950. *The I Ching or Book of Changes*. Translated by Cary F. Baynes. Princeton: Princeton University Press, Bollingen Series XIX.

_____, and C. G. Jung. 1962 [1929]. *The Secret of the Golden Flower*. New York: Harcourt, Brace and World.

Zhao Juan 赵娟. 2011. "Wenti yu shijiao: Xifang yixue yanjiu de sanzhong lujing" 问题与视角：西方易学研究的三种路径. *Zhouyi yanjiu* 周易研究 2011/4.

Zhouyi 周易. 2010. "Hanxue shi yezhong Wei shi fuzi de Zhouyi yijie yu yanjiu" 汉学视野中卫氏父子的周易译介与研究. *Zhouyi yanjiu* 周易研究 2010/4.

Immortality and Meaning in Life

A Daoist Perspective

JEFFREY W. DIPPMANN

> Essentially, the problem with immortality seems to be one of inevitable boredom. The problem is tedium. . . [In the end, the] seemingly positive dream of immortality becomes a nightmare, a nightmare from which we can never escape.
> — Shelly Kagan, *Death*

> Master Whitestone . . . was already more than two thousand years old. . .[and] Pengzu asked him, "Why do you not ingest drugs that would enable you to ascend to the heavens?"
>
> Master Whitestone replied, "Can one amuse oneself on high in the heavens more than in the human realm? I wish only to avoid growing old and dying."
> — Ge Hong, *Shenxian zhuan*

What if we could attain a physically immortal existence? Would it not be desirable to do so, to extend the amount of time we have to pursue all of our life goals, to amuse oneself on earth, to finally "have time enough at last"?[1]

The philosophical and religious quest for the meaning of, or meaning in, life has produced lively discussions on the feasibility and desirability of an immortal existence and its relationship to time. On the surface, the question of its appeal would seem to be nonsensical. Who would not want to live forever? Are we not all seeking to extend our lives in one manner or another, fearing and assiduously avoiding that moment when our conscious life is no more, and we face the potential abyss of non-existence?

The promise (possibility?) of immortality, in a myriad of different guises, has animated human speculation and literary imaginations across the world and throughout history. One simply need look to works such as the Babylonian epic *Gilgamesh*, Greek mythology (in particular the story of

[1] The title of a classic 1959 *Twilight Zone* television episode.

Tithonus),[2] philosophical treatises from Plato's *Phaedo* to Moses Mendelssohn's *Phaedon,* early modern gothic literature (see Mulvey-Roberts 2016; Lien 2018), Jonathan Swift's 1726 classic *Gulliver's Travels,*[3] and elsewhere to see the prevalence and sustained interest in the subject.

Indeed, even a cursory search for the theme in contemporary literature reveals a resurgence in works that either focus exclusively on the nature of immortality (e.g., Robbins 1990; Magary 20111 Cross 2014) or include immortals as prominent characters. And while China has a rich history of hagiographic and literary accounts of immortals (*xian* 仙), that we will explore below, we are also witnessing an upsurge in that nation's online serials and print works reintroducing the figure to contemporary Chinese audiences.

Aside from the question of the soul (which is outside the scope of the present study), most Western thinkers have focused on the issue of bodily immortality and almost unanimously reject the idea as either untenable or undesirable.[4] At stake is the question of the tolerability of endless time in physical form: could we actually bear to live on earth for 300, 500, 1500 years or more? Numerous objections have been raised to such a prospect, including issues involving the loss of loved ones,[5] the raw fact of our own death infusing life with meaning,[6] the "unnatural" or "evil" character of

---

[2] Tithonus' story is one of the earliest cautionary tales about the inadvisability of immortality, in that he was granted eternal life by Zeus, but not eternal youth. Upon reaching old age, he was secluded in a room by his lover Eos, where he spent the remainder of his time endlessly babbling. A modern take on this myth can be found in the X-Files television series ("Tithonus," season 6, episode 10, 1999).

[3] Part III, chapter X has the story of the *struldbrugs,* an immortal but ever-aging people living on the fictional island of Lugnagg. Like Tithonus, while possessing immortality, they nevertheless undergo the ravages of increasing infirmity, loss of memory, and "deformities." Interestingly, Lugnagg is geographically located southeast of Japan and may have been Swift's nod to the Chinese land of immortals, Penglai.

[4] John Martin Fischer identifies these thinkers as "immortality curmudgeons" (2020, 89). I couldn't have come up with a better description!

[5] See Mary Shelley's 1833 short story "Mortal Immortal," in which the protagonist Winzy despairs of his loneliness as one after another of his loves dies, leaving him with "no beacon except the hope of death," although he alone among all mortals has become bereft of the "sheltering fold" of death. The X-Files episode referenced above also touches on this concern. In it, the Tithonus-like character bemoans the fact that not only has he lost his wife, but needed to return to the records office simply to recall her very name.

[6] Among others, we can cite Viktor Frankl, who argued "What would our lives be like if they were not finite in time, but infinite? If we were immortal, we would

immortality (e.g., Lien 2018), and more. This chapter's scope, however, is limited and not intended to answer such questions.

Instead, the present chapter examines the issue from the question of boredom, as famously discussed in the Oxford philosopher Bernard Williams' essay, "The Makropolus Case: Reflections on the Tedium of Immortality" (1973). Animating much of the debate over the past half-century, his principal thesis revolves around the argument that an immortal life would be crushingly boring, given the vast expanse of time involved and inevitable repeatability of our experiences. Along with its relevancy in terms of perspectives on time, the question has increasingly found its way into popular and scientific discourse both as a growing condition of modern life and our navigation through the past twelve-plus months of the Covid-19 pandemic.[7]

Indeed, new studies regularly appear in the catalogs of both university and mainstream presses, including the first collection of original essays devoted to the topic, the *Boredom Studies Reader* which says on the cover,

> Boredom Studies is an increasingly rich and vital area of contemporary research that examines the experience of boredom as an important—even quintessential—condition of modern life. . . [and this] book considers boredom as a mass response to the atrophy of experience characteristic of a highly mechanised and urbanised social life. (Gardiner and Heladyn 2016)

As *The New Yorker* columnist Margaret Talbot puts it, "In the past couple of decades, a whole field of boredom studies has flourished, complete with conferences, seminars, symposiums, workshops, and a succession of papers [in academic journals]" (2020).

The present chapter addresses the question from one perspective in the Daoist tradition and its generalized goal of becoming a *xian*, immortal or transcendent, especially as portrayed in the 4th-century work *Shenxian zhuan* 神仙傳 (Traditions of Divine Transcendents; trl. Campany 2002) by

---

legitimately postpone every action forever. It would be of no consequence whether or not we did a thing now; every act might just as well be done tomorrow or the day after or a year from now or ten years hence. But in the face of death as absolute finis to our future boundary to our possibilities, we are under the imperative of utilizing our lifetimes to utmost, not letting the singular opportunities- whose 'finite' sum constitutes the whole of life-pass by unused" (1957, 73).

[7] Paul Thagard describes COVID-19 fatigue as "a complex of emotions that include boredom, loneliness, sadness, frustration, anxiety, fear, anger, and resentment, all brought on by the loss of activities and social relations produced by pandemic restrictions" (2020).

Ge Hong 葛洪 (283-343 CE). We will not enter into the debate concerning the ontological "reality" or plausibility of immortality, nor whether the immortality pursued by Daoists is "physical" or "spiritual." Instead, irrespective of the veracity of Daoist claims to transcendence or immortality, the prevalence of the literary motif and its continued pursuit by adepts lend themselves to an examination of why immortality is an acceptable and valued goal for Daoist practitioners.

I am neither interested in establishing that all Daoists at all times have pursued a literal, physical immortality nor in establishing that such immortality is philosophically rational or plausible. Contemporary scholarship acknowledges that the tradition's history has embraced a wide variety of beliefs, attitudes, and practices concerning the nature of death, and any attempt to hint at an orthodox position is doomed at the outset. However, the literary trope, the hagiographic motif, has demonstrably influenced Chinese culture, ranging from short stories, novels, and poetry, all the way to contemporary television and film. Immortality cultivation (*xiuzhen* 修真) themes, moreover, have resurged in contemporary Chinese science fiction. Based on these, we will explore perceptions and outlooks rather than realism.

Kenneth DeWoskin points to the cultural impact of hagiographic literature when he writes that "*xian* are described not as they actually are, but in such a way as to make them conceivable to readers, who are mostly mortals, interested in time, mortality, and the remaining gamut of human issues" (1990, 79). In making the ideal "conceivable," the stories provide us with models of what is possible, what we can strive for, who we can potentially become. To do otherwise, to lay before the reader an unattainable or "unthinkable" future, one with no connection to their present circumstances or aspirations, would render the work of little sustainable worth.

While arguing against the position that adepts literally seek physical immortality, Fabrizio Pregadio also acknowledges the role that such "tales" have played in China, helping to create "an image of Daoism that was easily understood by persons of different education and background, and that was also tolerable by followers of other intellectual or religious traditions: after all, the tales about the immortals are only tales" (2018, 385). As Chi-tim Lai notes, the literary "symbolic world of *xian*-immortality provides an individual ideal that may look fantastic and literally impossible, yet, the symbolic *xian*-images are no less useful and constructive in giving meanings to . . . life . . . and personal ideal identity" (1998, 209). Or, as Russell Kirkland puts it, the immortality model "provided fuel for centuries of both religious and literary elaboration. . . [while the term *xian* came] to denote an exemplar of spiritual qualities on a level sufficient to allow a transcendence of human mortality" (2004, 185).

The Makropulos Case

Before looking at the nature of boredom, a summary of Bernard Williams' essay is in order. Based on a lecture at the University of California, Berkeley, he published "The Makropulos Case: Reflections on the Tedium of Immortality," in his *Problems of the Self* in 1973. Since then, the essay has appeared in numerous philosophical anthologies and been the subject of dozens of articles both supporting and rejecting his basic premises.

As he states in his opening, his principal thesis is that "Immortality, or a state without death, would be meaningless . . . so, in a sense, death gives meaning to life" (2010, 345). His starting point is the fictional case of Elina Makropulos, a character in a play by the Czech writer Karel Čapek (1922), in 1925 adapted as an opera by Leoš Janáček.[8] Elina Makropulos, at the time of the action, is 337 years old: she has been 37 for 300 years. Her extraordinary longevity has come about through ingesting an elixir of life developed by her father, a 16th-century physician.

Having taken the elixir, she has apparently remained in the physical condition typical of a vigorous, robustly healthy 37-year-old. She appears to have no particular reason to doubt that this state will continue, provided she keeps taking the elixir every three hundred years. It quickly becomes evident, however, that Elina's mental state is one of great suffering: after three hundred years, her potentially unending life has become unendurable to her. Caring about nothing, finding value in nothing, incapable of recalling the number of children she has borne, she retains the instincts allowing her to move along from one day to the next, but those rudimentary impulses are now alien to her since they preserve nothing she cares about. In describing her existence at 337, she plaintively laments, "Everything is irksome. It is tiresome to be bad and tiresome to be good. Heaven and earth tire one. And then you find out that there truly is none. Nothing exists—neither sin, nor pain, nor desire—absolutely nothing."

---

[8] The episode to which Williams refers appears in the play's third act. Discrepancies about the age of Elina Makropulos exist between the play and Williams' recounting. Burley helpfully locates these in the fact that the original play identified EM as a woman of 337, while the operatic version that Williams was apparently familiar with placed her age at 342 (2009, 306). We will use the former throughout this chapter.

According to Williams, "her unending life has come to a state of boredom, indifference, and coldness. Everything is joyless . . . an endless life was meaningless" (2010, 346), the "tedium of immortality" in the essay's subtitle. Ultimately refusing to take another dose of the elixir, she dies, and the secret formula is destroyed by a young woman. From Williams' perspective, this is the right and proper choice, for immortality would be unbearable. Such an existence would be intolerable for a human being, insofar as interminable boredom would eventually set in, along with despair.

Yale philosopher Shelly Kagan aligns himself with Williams' basic premises. He writes, "I find myself agreeing with Bernard Williams when he says that immortality wouldn't be desirable. It would actually be a nightmare, something you would long to free yourself from" (2021). Among other considerations, he laments what he perceives to be the seemingly endless repetition of an immortal existence:

> I have a friend who once claimed to me that he wanted to live forever so that he could have Thai food every day for the rest of eternity. I like Thai food just fine, but the prospect of having Thai food every day, day after day, for thousands, millions, billions, trillions of years does not seem to me an attractive proposal. It seems to me, rather, a kind of a nightmare. . .
>
> Essentially, the problem with immortality seems to be one of inevitable boredom. The problem is tedium. You get tired of doing math after a while. After a hundred years, a thousand years, a million years, whatever it is, eventually you are going to say, "Yes, here's a math problem I haven't solved before, but so what? I've just done so much math, it holds no appeal for me anymore."
>
> Or you go through all the great art museums in the world (or the galaxy) and you say, "Yes, I've seen dozens of Picassos. I've seen Rembrandts and Van Goghs, and more. I've seen thousands, millions, billions of incredible works of art. I've gotten what there is to get out of them. Isn't there anything new?" And the problem is that there isn't. There are, of course, things that you haven't seen before—but they are not new in a way that can still engage you afresh. (2012, 243)

From his bird's eye view, Kagan appears to envision immortality as a succession of repeated events and activities, experiences that do not change over the course of time, thereby becoming redundant and meaningless. For meaning to be present, there must be something "new" in those experiences; in other words, change must have occurred over the course of time. Without change, without fresh engagement, there is nothing for me "to get out of them."

Boredom and Time

In contrast to its recent popularity, boredom as a subject has traditionally received little attention among philosophers and scientists. While elements of the concept can be traced back to Greek philosophy and medieval monasticism, the emergence of "boredom" as we now think of it appears to have its roots in the modern, urban mechanized age (see Spacks 1995; Swendsen 2005). As many have argued, the increasing possibility and expectation for leisure time among urbanites (at least among the elite), and the sharp demarcation for laborers between highly structured work and open non-work hours has contributed to a sense of free, unoccupied time, and hence a sense of what Dickens first identified as "boredom" in his 1852 novel *Bleak House*. However, an exhaustive definition and analysis of boredom's nature lies outside our immediate concerns. Instead, for our purposes, a few salient observations about time's relation to boredom and its possible causes will suffice.

From the perspective of "folk wisdom," we all know that time drags and strikes us as interminably long while in the throes of boredom, in sharp contrast to how "time flies when you're having fun." Who among us hasn't lamented the slow march of clock hands as we sit through uninteresting shows, lectures or the increasingly ubiquitous Zoom meetings? Charles Simic aptly captures its subjective feeling in the opening stanza of his poem "To Boredom":

> I'm the child of your rainy Sundays.
> I watched time crawl
> Over the ceiling
> Like a wounded fly. (2007)

Martin Heidegger, who wrote the most extensive exploration on boredom until recent times, derived from a series of lectures delivered 1929-1930, inextricably tied the subjective sense of boredom to time. In the following passage, his starting point is the literal meaning of the German word routinely translated as boredom, *Langweile*, long time or long while:

> Boredom, *Langeweile*—whatever its ultimate essence may be—shows, particularly in our German word, an almost obvious *relation to time*, a way in which we stand with respect to time, a feeling of time. Boredom and the question of boredom thus lead us to the problem of time. We must first let ourselves enter the problems of time, in order to determine boredom as a particular relation to it. Or is it the other way around, does boredom first lead us to time. . . . Or are we failing to ask correctly concerning either the

first relation—that of boredom to time—or the second—that of time to boredom? (1995, 80)

William James proposed that this occurs because segments of time

> . . . shorten in passing whenever we are so fully occupied with their content as not to note the actual time itself. A day full of excitement, with no pause, is said to pass 'ere we know it.' On the contrary, a day full of waiting, of unsatisfied desire for change, will seem a small eternity. *Tedium, ennui, Langweile, boredom,* are words for which, probably, every language known to man has its equivalent. It comes about whenever, from the relative emptiness of content of a tract of time, we grow attentive to the passage of the time itself. (1890, 626)

This points to two predominant causes at the root of our experience of boredom: change and desire. James focuses on our unrequited longing for change. Whether due to some innate quality of human nature or the emergence of the modern age, our need for change and newness appears fairly deep rooted. While professing that we simply want stability in our life, a settled place that we can call home, a nostalgia for what once was, the reality is that we often unreflectively become restless and subsequently bored when faced with an unchanging present or future.

Recall Kagan's lament about the lack of freshness he envisioned during an extended period of time. Again, the past year of Covid-19 has subjected many to a life of social distancing, confined to their homes, and lacking clearly defined borders between days and experiences. As a result, many individuals struggle with Covid fatigue and boredom. Without the change in environment, activities, and attainable goals, people worldwide report being disconcerted and adrift. Is today Monday, Wednesday, or Saturday? Is it March or January, 2021 or 2020? Change allows us to demarcate time, thereby placing meaning and value on our activities (see Gontcharova 2021; Wittmann 2020).

Our desire for change paradoxically runs parallel to our longing for stability. Leo Tolstoy famously depicted boredom as the "desire for desires" (Tolstoy 2004). Adam Phillips, a British psychoanalyst, examined the phenomenon in both children and adults, and concluded:

> We can think of boredom as a defense against waiting, which is, at one remove, an acknowledgement of the possibility of desire. . . In boredom . . . there are two assumptions, two impossible options: there is something I desire, and there is nothing I desire. . . In boredom there is the lure of a possible object of desire, and the lure of the escape from desire, of its meaninglessness. (1993, 76)

In his 1930 work, *The Conquest of Happiness*, Bertrand Russell devotes chapter 4 to "Boredom and Excitement." Here he posits that boredom is a distinctively human emotion, "essentially a thwarted desire for events, not necessarily pleasant ones, but just occurrences such as will enable the victim of *ennui* [boredom] to know one day from another" (1930, 60). Shades of Covid-19 restrictions!

And of course, we only desire that which we do not already have, thereby necessitating change—the movement from our current state/situation to a new one in which the object of our desire has been attained. Once Heidegger's train has arrived, the pandemic has abated, the Zoom meeting ends. . . our desire is satiated—for the time being. As Tolstoy observed, we desire to desire. Our lack of contentment, our inability to accept the present as it is, and not how we want it to be, lies at the root of both positively animating ambition and stultifying boredom.

Goals and Desires

Bernard Williams, then, posts that an acceptable scenario for immortality must meet two fundamental criteria: first, the future person must be genuinely identical to the present individual, not simply qualitatively similar or having several identical properties; and second, the person's future life must be attractive and pleasurable, in line with his or her goals.

The latter is the more problematic for Williams in that, if our goals remain the same throughout eternity, ultimately boredom, fatigue, or detachment will occur. This is "a boredom connected with the fact that everything that could happen and makes sense to a particular human being of forty-two had already happened to her" (2010, 352). And if goals have changed, can we then assume it is the same identical individual? Williams says that it cannot.

> Any coherent condition making immortality palatable must provide a "model of an unending, supposedly satisfying, state or activity which would not rightly prove boring to anyone who remained conscious of himself and who had acquired a character, interests, tastes, and impatiences in the course of living, already, a finite life. . .
>
> Nothing less will do for eternity than something that makes boredom unthinkable . . . [perhaps] guaranteed to be at every moment utterly absorbing." (2010, 356-57)

What gives life meaning is the activity of pursuing one's deep, categorical desires (Williams 2010, 348). Categorical desires are not conditioned by, or dependent on, one's continued existence but constitute a compelling reason to go on living. Such desires serve to propel one into the future. For example, the parental desire to care for young children until they can care for themselves can be categorical by providing sufficient reason for wanting to continue living.

However, Williams also argues that categorical desires are ephemeral and exhaustible, either through their satisfaction or by losing interest in them. Additionally, the desire to attain and/or experience pleasure or happiness cannot be a categorical desire in his estimation. Instead, categorical desires are those that motivate a person to stay alive *despite* the prospect of unpleasant times (2010, 354). People who lead a life full of suffering would not be motivated to stay alive by a desire for pleasure or happiness if they only saw unending suffering for the future. The only way to be immortal and not exhaust categorical desires is to have an infinite number of disjointed lives, and not an extension of the same person—but this would violate Williams' first criterion, that the future person must be genuinely identical to the present.

If we seek to mitigate that fate, thereby avoiding the inevitably of boredom, we could acquire entirely new categorical desires whenever the old ones were exhausted. However, this option violates Williams' second criterion, that over time the projected future-self eventually will not pursue the same single project, dream, or goal he or she presently cares about. This contingency makes it even more difficult to justify the claim that immortality could be desirable: why would we wish to be immortal if that endless life is only pursued in order to complete projects, dreams, and goals that we currently don't even know or care about? (Williams 2010, 357; see also Kagan 2012).

Therefore, according to his analysis, "an endless life would be a meaningless one . . . We could have no reason for living eternally a human life. There is no desirable or significant property which life would have more of, or have more unqualifiedly, if we lasted forever" (2010, 357). Such a life would lack any substantial meaning, due to the interminable repetition of activities and experiences predicated on the sameness of goals, categorical desires, etc. Samuel Scheffler agrees: "[When categorical desires die, which they must] . . . one will be left with nothing *but* oneself, and one will be doomed to a kind of boredom from which there is no chance of escape in this world. . . we will be left *with* ourselves, and we ourselves are, terminally boring" (2015, 94-95). A gloomy prospect indeed, at least from this philosophical perspective.

Challenges

The viewpoints expressed by Williams, Scheffler, and Kagan have not gone unchallenged. Dozens of essays have appeared over the past fifty years both supporting and critiquing their analysis. However, there are two that have not been addressed and that also serve to introduce the Daoist alternative.

The first centers on the bird's eye view of immortality. Of necessity, speculations about the prospects of immortality are just that, thought experiments loosely based on philosophical and psychological principles. Actual accounts of immortals, both Eastern and Western, are nothing more than tales as opposed to first-hand observations (accepting Pregadio's characterization). The perspective, therefore, is founded on what can be imagined or thought about. And for this exercise, Western philosophy has taken what I view as a grand, 50,000-foot-high perspective.

Recall Shelly Kagan's views on the relentless dreariness and monotony of an eternal life, one that would presumably be an immortality with no fear of death—either accidental or medical, such as envisioned in Daoist tales. In each instance he cites, it is clear that the source derived from his projection of what immortality might be like. And for this, he engaged in a thought exercise, through which he could survey a vast expanse of time from "above," as did Williams, Scheffler, and others. He writes,

> Humans have this ability to look down on their experiences, or to step back from their experiences, and assess them. Even now, for example, as I'm sitting here typing these words, watching the screen on my computer, listening to the birds outside my window, part of me is thinking about whether I am getting my point across, and whether the light coming in from my window is a bit bright, and so forth and so on.
>
> We can all reflect on our first-order or base-level experiences, even while we are having them . . . [In assessing your future] You're going to take this meta-level or higher-level standpoint, look down . . . and wonder, "Is this all that there is to life?" (2012, 243)

The problem with the view from the heights, as I see it, is that our perception of what is below us, or stretching in front of us in this instance, devolves into a kind of sameness that blurs the details making the particular what it is. Your view of New York City from an airplane at 10,000 feet bears little resemblance to the perspective you get while standing in Times Square. The identity of the "river" that you are about to cross changes dramatically as you consider the fact that there is no such "river" in front of you, only a chain of individual water droplets. I will always recall my first week in Ellensburg, WA, when a local artist assured my wife and me that the dull brown ridges encircling the city were in actuality very beautiful combinations of variegated vegetation, increasingly visible the closer we got.

Accordingly, I think that when we "step back . . . [and] reflect on our first-order or base-level experiences," as well as survey 10,000-plus years, it is all going to look very much the same. The monotony, the tedium, the repetition identified by Kagan, Williams, and others strikes me as having that same quality, a blurring together of the individual moments, experiences, etc. that make up our daily life. Of course, eternal life will project out as one continuous dull experience when envisioned in the abstract. But does it indeed look that way from the ground and in the moment?

Such an accounting does not take into consideration the inevitably of change. While we often gloss over the subtle (and not so subtle) alterations in phenomena that occur with regularity, they nevertheless are present in each moment and continuously render the world anew to the keen observer. Kagan's examples of Thai food, math problems, great works of art, etc. are all portrayed as static phenomena. Once experienced multiple times and with the potential to continue *ad infinitum*, they lose their appeal since there no longer exists anything fresh or new that can engage us.

However, if, as the Daoist worldview proposes, things are in constant flux, alternating between yin and yang, flowing through the five phases, imperceptibly changing from moment to moment, day to day, month to month, isn't it just as plausible that we *can* discover something "new under the sun" if we simply pay heed? Close attention to today's pad thai, even if we made it just like last week's pad thai, yields subtle differences brought about by the inevitably of change.

One of my favorite pieces of music is Ralph Vaughan Williams' "Fantasia on a Theme by Thomas Tallis," performed by Sir Neville Marriner and the Academy of St. Martin-in-the-Fields orchestra. Note first that Williams created a new composition based on an existing, older one. Secondly, I have not only grown in my appreciation for that particular recording over the past thirty-plus years, but have also discovered comparable beauty in the interpretations of other conductors and orchestras. I have no reason to doubt that I will have similar experiences over time and/or that new variations on Williams' "Fantasia" will appear.

In contrast to the view from above, the attentive view from the ground and in the moment can yield fresh experiences and engagements due to the constancy of change. I will grant that we rarely stop to take note of those details, and would argue that therein lies our present state of boredom, a state overcome by Daoist immortals.

The second challenge relates to the context, within which we moderns find ourselves: a condition of separation from the rhythms present in, and engaged with, nature. Urbanization, modernization, and mechanization have all contributed to an overwhelming desire for distraction and a fast-paced life that leaves little time for boredom to set in. In addition, the creation of cities has inevitably taken away from our ability to remain part of the natural world: we often find ourselves confined to "concrete jungles," intentionally designed to look the same. Bertrand Russell identified the malaise of contemporary life when he observed that, "the special kind of boredom from which modern urban populations suffer is intimately bound up with their separation from the life of the Earth" (1930, 67-68).

After once witnessing the "ecstasy" of a two-year-boy allowed to leave London for his very first time, joyously burying his face in the wet, muddy earth of the country, Russell reflected that

> whatever we may wish to think, we are creatures of Earth; our life is part of the life of the Earth, and we draw our nourishment from it just as the plants and animals do. . .
>
> Such pleasures. . . that bring us into contact with the life of the Earth have something in them profoundly satisfying. . . The two-year-old boy . . . displayed the most primitive possible form of union with the life of the Earth. (1930, 66-67)

A generation earlier, the American Transcendentalist Henry David Thoreau, while living simply at Walden Pond, came to much the same conclusion. There, he found that without the distractions of the city, newspapers, factory bells, and the ticking hands of the clock, nature had a rhythm of its own. Morning brought with it a continuous renewal, an awakening hour that lent itself to an appreciation of the "faint hum of a mosquito" and an opportunity to awaken and "live deliberately" with full attention to life as it unfolded.

Desiring and pursuing the distractions of contemporary life only leads to a slumber, from which most people cannot (or do not want to) awaken, a "drowsiness" that robs of us of our capacity to truly be "alive." This slumber is the boredom of those who "give so poor an account of their day." We therefore must awake, for "to be awake is to be alive." In place of the artificial, superfluous activities of those sleepwalking through life, Thoreau advocates that we "spend one day as deliberately as Nature. . . settle ourselves . . . till we come to a hard bottom and rocks in place, which we can call *reality*, and say, 'This is, and no mistake'" (1992, "Where I Lived, and What I Lived For"). Little can be gained from a life hurried, in constant pursuit of desires that always remain out of reach, waiting for that which is just down the road, tomorrow.

> Time is but a stream I go a-fishing in. I drink at it; but while I drink, I see the sandy bottom and detect how shallow it is. Its thin current slides away, but eternity remains. . . I do not wish to be any more busy with my hands than is necessary. (1992)

To be awake is to take the full measure of our time here, to note the hum of the mosquito's wings, the sandy bottom of the shallow stream, the reality of simplicity and the flow of untrammeled nature. To apprehend this truth, and our place within it, is to be alive, to have rediscovered our roots, and ultimately to find contentment therein.[9]

In the end, it appears to me, the boredom of which Williams and others write is that of modern urban civilization, the tedious nature of sedentary work, and the seemingly unchanging sameness of our average day/week/month, especially exaggerated during the year of Covid-19. Added to that disconnect are all the distractions with which we are bombarded on a daily, nay momentary, basis. Smart phones, computers, DVR systems and any number of other modern "conveniences" demand our constant attention, alternately speeding up our perception of time as we flit from one to another, and slowing it down when we are forced to disengage from them. What appears to have been lost is our natural footing, increasingly becoming what Russell warned would be "a generation that cannot endure boredom . . . a generation of little men . . . unduly divorced from the slow processes of nature (1930, 65).

As he notes, however, a life at one with and in nature can be profoundly satisfying and stimulating, as we return to the roots of our being and connection with the world around us. And this is exactly the vision brought forth in depictions of the Daoist immortal, especially the "earthbound." For those choosing such an existence, life is continuously wondrous, not something one wishes to escape, either through death or even celestial transcendence.

---

[9] For yet another compelling vision see William Wordsworth's *Ode on Intimations of Immortality from Reflections on Early Childhood* (2011) and its analysis in relation to boredom (Thompson 2019). Interesting analogies can be drawn between Thompson's insights into God as the ground of our being and the Daoist perspective on Dao as ultimate root.

The Daoist *Xian*

Considerable debate continues as to the exact nature of the *xian* (translated here as immortal), and whether they attain bodily immortality. It is clear, however, that one goal for Daoist adepts involves the attainment of some sort of immortality. Traditionally, three types of *xian* appear in the literature: celestial immortals (*tianxian* 天仙) who have already ascended into heaven and occupy a position in the heavenly bureaucracy; earth immortals (*dixian* 地仙), ready for ascension who yet decide to remain on earth; and those who transcend this world through deliverance from the corpse (*shijie xian* 尸解仙), leaving behind a token or substitute upon ascension.

It is important to note that etymologically, the approbation *xian* has been linked to the combination of two free standing characters: *ren* 亻, person, and *shan* 山, mountain. Thus, leading Daoist scholars such as Kristofer Schipper to the interpretation that an immortal is "the 'human being of the mountain' or alternatively 'human mountain.' The two explanations are appropriate to these beings: they haunt the holy mountains while also embodying nature" (1993, 164).

The key figures of interest here are the earth immortals, primarily as depicted from the Han to the Eastern Jin dynasties (206 BCE-317 CE) but also as they appear in contemporary Chinese science fiction. First, they most closely resemble the Western characterization of immortality presented above. The "lives" of celestial immortals offer ideals that speak little about finding meaning in life here in the world. "Deciding" to ascend to heaven, leaving behind the mean world of an earthbound existence, more closely resembles the characteristics of an immortal, spiritual soul found in Western philosophical and religious traditions. And while some theoreticians, like Kagan, will expand their critique of immortality to an "afterlife" in heaven, most confine their arguments to the prospects of body bound existence (2012; 2021).

Second, earth immortals are in some sense the most important of the three types, both in classical and contemporary literature. While scholarship continues to address whether Daoist immortality was ever solely focused on the refinement of a spiritualized embryonic body (Pregadio 2018), there is little doubt that the period between the Han and Eastern Jin dynasties produced a great deal of literature celebrating the *daxian* and extolling the pursuit of earthly immortality.

Yü Ying-shih has observed that by the Han, China had undergone a dramatic change in its

> views on the life of *xian* immortals. In pre-Qin literature the *xian* is portrayed only as a secluded individual wandering in the sky, in no way related to the human world. But in Han literature we begin to find that the *xian* may sometimes also enjoy a settled life by bringing with him to paradise not only his family but also all chattels of his human life. (1964, 119)

Likewise, Zornica Kirkova in her recent study of medieval Chinese verse identifies a discernible shift in representation of immortals during the same period, one that marked a move from celestial to terrestrial figures (2016, chs. 4-5). Regardless of the reasons behind such a change—possibly related to the Han emphasis on social and familial conditions (Yü 1964, 119)—the period produced a major comprehensive hagiographic compilation of Daoist immortals, the *Shenxian zhuan*, which places earth immortals in a highly enviable and imitable position.

The collection's compiler was Ge Hong, also the author of the famous *Baopuzi* 抱朴子 ([Book of the] Master Who Embraces Simplicity), the earliest systematic exposition on immortality methods and defense of the practice. He offers a classification of immortals based on paths of cultivation and achievements:

> Superior practitioners who rise up in their bodies and ascend into the void are termed celestial transcendents (*tianxian*). Middle-level practitioners who wander among noted mountains are termed earth-bound transcendents (*dixian*). Lesser practitioners who first 'die' and then slough off are termed 'escape by means of-a-corpse-simulacrum transcendents' (*shijie xian*). (Campany 2002, 75)

However, while celestial immortals occupy the highest class, Ge Hong makes little secret that in his judgment, earthbound immortality is both acceptable and perhaps even preferable, given the fate of new [rookie?] immortals. While some ascend to heaven and others remain on earth, ultimately:

> What matters is that they have all achieved Fullness of Life, they simply make their abodes wherever they prefer. . .
>
> Once one's immortality has been confirmed, one is never again concerned about the fleeting of time. If one should return temporarily to wander on earth or in the famous mountains, what would there be to concern oneself about? Old Peng claimed that in heaven there were so many important gods holding offices of high honor that the newer genii [immortals] must hold the meaner positions. . . their lot is harder than before. He saw no point in his striving persistently to go to heaven, so he remained among men for eight hundred years and more. (Ware 1966, 64-65)

In an interesting side note, one highlighting the importance of eart immortals during this period, Isabelle Robinet speculates that the three immortal types can be viewed as "generally corresponding to the celestial, terrestrial, and underworld realms" (1997, 46). This insight references the notion of the *axis mundi*, the center post of the world present in many of the world's religious traditions, made famous in the pioneering work of Mircea Eliade (1959, ch. 1).

The *axis mundi*, or hub of the universe, represents the vertical connection between three sacred realms. Often symbolized by trees, poles, and mountains (such as Mt. Kunlun in Daoism, one major home of immortals), it serves as the locus where cosmic regions intersect and where the universe of ultimate reality is accessible in all its dimensions. The terrestrial axis is essential in bridging the gap between the celestial and underworld realms, acting as the primary conduit for communication and intercourse. It provides access to reality or being as it truly is. In fact, it can be viewed as defining reality, for it marks the place where being is most fully manifest. If Robinet is correct, then the terrestrial or earth immortal represents one of the most important living links in the Daoist universe.

Xinkai Huang alludes to a similar sensibility in online fantasy literature published during the 2000s. From among eight different categories used in the QIDIAN fantasy website system— the largest such site in the world, boasting ten million registered users in 2010 (Huang 2011, 120nn.2-3)—he chose to focus his research on the "Super-human perfection stories" where the goal is "to become immortal through exercising Daoism, Buddhism, or similar techniques" (2011, 123). His reason for doing so is because it is "the most popular fiction category and the selected stories were among the most popular ones from 2004-2008 in terms of reading visits and favorite-votes . . . The stories have many fans, who posted messages" (2011, 124).

Apart from illustrating the continuing Chinese interest in the pursuit of immortality, Huang's study also buttresses my suspicion that the terrestrial realm and its inhabitants (including those who aspire to become "super-humans") plays a critical role in the fantasy cosmology surrounding immortality. Hence, in *Untouchable Journey*, "the universe where average people are situated is the center that connects all higher universes" and serves as the place where the main protagonists return (2011, 125).

Additional support an also be presented for this idea. Stephen Bokenkamp, in drawing our attention to the different beings identified as immortals (such as earthbound, celestial, and cavern dwellers below us) in the Numinous Treasure (Lingbao) tradition, notes that they all share the quality of having been "transferred" from their "common human state to a more subtilized form of existence . . . There is thus not a single chasm between mortals and immortals, but a chain of being, extending from nonsentient forms of life that also experience growth and decay to the highest reaches of the empyrean" (1999, 22). Rather than being distinct, separate kinds of being, they exist on a continuum that links them together in three realms of Eliade's schema.

The close connection of the worlds, as much as the importance of the terrestrial realm as a communicative link between them, also appears in China's artistic tradition. One example is the 10th-century scroll *Langyuan nüxian* 閬苑女仙 (Goddesses in the Palace Park), reprinted and analyzed by Shih-shan Susan Huang (2012). Attributed to Ruan Gao 阮郜, the work depicts an immortal island frequented by goddesses. Highly interesting in its own right, of greatest importance for our purposes is the rock outcropping located in the scroll's lower right corner. "By connecting the isle to the land on the other side of the ocean and highlighting the isle's accessibility, the bridge-like rock formation invites and arouses further interest in the attainment of immortality" (Huang 2012, 111-13). With the right attention, with the proper practice, we can access a world of timeless immortality to which we are already connected.

For this reason, the hagiographic models of earth immortals have the most to say to aspirants, serving as powerful early counterpoints to the images drawn in Williams, Kagan, and Scheffler. Despite their transcendence of ordinary existence, attainment of seemingly super-human capabilities, and immunity to medical and accidental death, they are in the end humans like you and I. Numerous times Ge Hong assures us that immortals are not after all "a special species" (Ware 1966, 97). Instead, even "In the case of Old Peng [or perhaps better Laozi and Pengzu], . . . we are still dealing with mere men, not with creatures of a different species. It was through attaining the divine process that they enjoyed unique longevity, not through what they were by nature" (1966, 53). Even the legendary founder of Daoism, Laozi, "was someone who was indeed particularly advanced in his attainment of the Dao, but . . . not of another kind of being than we" (Campany 2002, 196).

Unlike the Biblical account in Genesis and its literary/theological descendants, the root of Daoist immortality lies not in a bargain with the devil or usurpation of divine prerogative. Immortality ultimately comes about by utilizing what humanity naturally possesses, what Ge Hong identifies as our "intelligence. If he can practice the same divine practices as did Old Peng, he can achieve the same results" (Ware 1966, 54). We must understand the process involved, put into practice specific methods and employ our innate qualities, but those are nothing more nor less than our mind: "Of all living things that have high intelligence, none surpasses man" (1966, 37). No need to appeal to heaven or the gods or anything beyond the earthly realm: "Our endowments are part of a natural process; they are not consignments from heaven and earth" (1966, 127).

Exemplars

To return to the observations of Bertrand Russell and Henry David Thoreau, perhaps a life led at one with and in nature can be profoundly satisfying, as we return to the roots of our being and connect with the world around us. By rejecting the artificiality of an increasingly modernized, "civilized" world constantly bombarding us with the urge to be entertained and kept engaged in that desired something lying "out there," just out of reach, thinking (or being convinced by advertisers perhaps) that we don't have it but certainly need it, can we avoid the stultifying boredom so vividly depicted by our "immortality curmudgeons"?

Slowing down to the pace of nature, recognizing the inevitability of change and the great potential it contains, seeing the granularity of pebbles within the stream of time, attentively listening to the hum of mosquito wings in the early morning—would these experiences not bring the continuous wonderment of Russell's two-year old boy burying his face in muddy grass for the first time? Rather than taking the philosopher's view from 50,000 feet or a million years, what would happen if we were to rest content in each discrete moment, desiring only that which is natural and immediately sufficient? Would an immortal life be as unthinkable and nightmarish as Williams, Kagan, and Scheffler imagine? The answer from the Daoist perspective is no.

The world as exciting, fresh and inviting is exactly the vision brought forth in many immortals' hagiographies, especially those associated with earth immortality. While the mortal world of this earth presents its challenges, it nevertheless contains everything that can constitute a satisfying and fulfilled life.[10] For those choosing such an existence, life is continuously filled with wonder, far from something that one wishes to escape, either through death or even celestial transcendence.

A first example comes from the *Shenxian zhuan* and recounts the hagiography of Master Whitestone (Baishi xiansheng 百石先生). He was already over 2000 years old at the time of Pengzu, another master of re-

---

[10] For example, the late 5th-century compilation *Zhen'gao* 真誥 (Declarations of the Perfected tells of the goddess Elühua, banished to "the world of stench and impurity" by the "Mystical Isles" for unexpiated sins (Bokenkamp 2021, 103).

nowned longevity (Campany 2002, 172). Living a simple life in the mountains, Whitestone subsisted on white stones, thus his name, and even took meat, alcohol, and grains—proscribed in many Daoist regimens. "He was unwilling to cultivate the way of ascending to transcendence, instead opting for nondying only, thus retaining the pleasures of the human realm" (2002, 292).

Once he encountered Pengzu, who asked him why he did not choose to leave the earth and ascend as a celestial. Master Whitestone replied: "Can one amuse oneself on high in the heavens more than in the human realm? I wish only to avoid growing old and dying. In the heavens above there are many venerable ones to be honored, and to serve them would be harder than to remain in the human realm" (2002, 294). Two thousand years in, an unbearable amount of time according to our curmudgeons, and Whitestone still took greater delight and found more pleasure in the world of humans than that of celestials.

In a similar vein, Ge Hong's *Baopuzi* speaks of venerable old masters:

> Anqi, Lord Nong of Longmei, Lord Xiuyang, and Yin Changsheng each took half doses of Potable Gold. They remained in the world for almost a thousand years, and only then did they leave it . . .
>
> Those who seek Fullness of Life merely do not wish to relinquish the objects of their current desires. Fundamentally, they are not yet overcome with any yearning to mount into the void, nor are they convinced that flying is always superior to being earthbound. If, by some good fortune, they can become immortal and yet go on living at home, why should they seek to mount speedily into heaven? (3.8a-8b, Ware 1966, 65-66)

The wise border guard in the *Zhuangzi*, who upbraided Yao for the latter's rejection of "long life, riches, and many sons," also points to an enviable existence on earth: "After a thousand years, should he weary of the world, he will leave and ascend to the immortals" (ch. 12, Watson 1968, 130). As opposed to a perceived life of tedium, boredom, and lethargy after 337 years, the Daoist perspective imagines that perhaps, "should" the sage grow tired of the world, he or she would ascend to join the celestials. But that leave-taking is far from a certainty and not driven by the nightmarish vision imagined by Kagan and others.

Ge Hong's collection also presents a snapshot from the life of Master Horseneigh (Maming sheng 馬鳴生), a disciple of Master Anqi 安期. Despite having received alchemical scriptures and refined a medicine that would allow him to rise to the celestial bureaus,

> Horseneigh took no delight in ascending to heaven but preferred to become an earthbound transcendent. He traveled about through the nine provinces for over five hundred years, no one realizing that he was a transcendent, as he built himself a house and raised animals like ordinary people. . .
>
> People did wonder at his nonaging however. In the end, Horseneigh did choose to leave the earth and join the ranks of the *tianxian*. (Campany 2002, 324)

Returning to Pengzu, one of our curmudgeonly arguments is addressed eloquently in his hagiography. Recall Mary Shelley's protagonist (see note 5 above) who bewails his immortal lot as those around him die. In the Daoist narrative, Pengzu does not fall into despair as his life progresses. Instead, he reflects:

> When my age had passed one hundred, I suffered some affliction. I have buried forty-nine wives and have lost fifty-four sons, and thus many times have I encountered trouble. The harmony of my pneumas was broken and injured, and through cold and heat my tissues have not remained vibrant; through my ups and downs I have dried up and withered. I fear I will not transcend the world. (Campany 2002, 176)

And yet, nowhere in his soliloquy with an inquirer does he indicate that he was disappointed or bored with his life. Instead, the conversation, and record as a whole, concludes with Pengzu transmitting his practices (and purportedly scriptures) to those wishing to follow his path. We can also contrast this with Čapek's (and by extension Williams') Elina Makropulos, who admits that she neither cares for the well-being of her children nor indeed can she remember how many she bore, "About twenty, I think. One loses count."

In numerous Daoist narratives, immortals do not seek to remove themselves from society, either to avoid relational complications or abandon the world of desires altogether. Granted, China's historical and cultural landscape is dotted with periods and individuals pursuing a life of reclusion as a path to immortality (Berkowitz 2000, 47-63). However, a recurring theme (and one designed to appeal to the popular imagination), focuses on the continued familial and social engagement of earth immortals. To reiterate Yü's observation, *xian* were often depicted as enjoying or extending "a settled life by bringing with him to paradise not only his family but also all chattels of his human life" (1964, 199). Accordingly, both Ge Hong and his hagiographic subject Pengzu criticize those who abandon the world in a misguided attempt to attain *xian*-hood. Ge Hong cites his own teacher in rejecting reclusion:

> To be immune from cold, heat, wind, and wet; invulnerable to ghosts, gods, and demons; unaffected by weapons and poisons; and never to be involved in the toils of grief, joy, slander, or praise: these are honorable. To turn one's back upon wife and children and make one's abode in the mountains or marshes, uncaringly to reject basic human usage and, clodlike, to become a companion of trees and rocks, is hardly to be encouraged. (*Baopuzi* 3.8a, Ware 1966, 65)

Likewise, Pengzu compares the shortcomings of those seeking a secluded life with the praise due to those who have truly entered the Way:

> Although these sorts [those who seclude themselves] have deathless longevity, they absent themselves from human feelings and distance themselves from honor and pleasure. There is that in them which resembles a sparrow or a pheasant transmuting into a mollusk: they have lost their own true identity, exchanging it for an alien pneuma. With my stupid heart I cannot bring myself to desire this.
>
> Those who have entered the Way should be able to eat tasty food, wear decent clothes, have sex, and hold office. Their bones and sinews firm, their complexion smooth, they grow old but do not physically age; extending their years, they long remain in the world. Cold, heat, aridity, and humidity cannot injure them; ghosts, spirits, and sprites do not dare attack them. Neither weapons nor noxious creatures can approach them; anger, joy, failure, and fame cannot entangle them, yet it is they who are worthy of esteem. (Campany 2002, 177)

Many more examples could be cited of immortals not only achieving their own goals, but also willingly bringing their wives, children, and even livestock with them. Again, rather than being a burden or a distraction, elements of the human realm and earthly life were embraced as meaningful and to be cherished.

Comparisons and Conclusions

Why this difference in attitude/outlook/perspective between Western curmudgeons and Daoist hagiographies? To me, it is due to the Daoist belief that all of life is interconnected, permeated in Dao thorugh the vital force of *qi*. We are not in any substantive metaphysical manner separate from the life of the earth and the realm of humans. As documented in the *Daode jing, Guanzi, Zhuangzi, Taiping jing*, and Ge Hong's works, the fundamental understanding in Chinese cosmology is that all existence is constituted by *qi*. In nearly all accounts, the loss of *qi* is the direct progenitor of death. Our connection with, and reliance on, *qi* in harmony with heaven and earth is the key to a long and satisfying life.

Our natural place is found right here and right now. Life, when properly understood, is good and enjoyable, worthy of extension through incalculable stretches of time. Our connectivity, is summed up best by James Miller in his work, *China's Green Religion*. He characterizes the Daoist perspective on the human place in the world as one of "insistence" rather than "existence," acknowledging people's interconnectivity with "what we moderns term nature, and also what we moderns term gods, not as beings or spaces that exist beyond the realm of human bodies, as it were, 'out there, but rather as subjectivities that 'insist' or 'dwell within' the space of the body" (2017, 12).

If the world "insists" within us, and we by extension "insist" in the world, then everything we could possibly require or need is already present. Our deluded (dare I say, arrogant) position of existence in the world, a world separate and distinctive from ourselves, ultimately fuels our desire for more—more of what we presumably lack. Again, we only desire that which we do not already possess. Projecting ourselves into the future, anxiously awaiting that which is yet to come (Heidegger's train, Kagan's fresh engaging experience), is both driven by our "desire for desires" and lies at the root of our boredom. Change and eternity (the ever present now) are already beneath our feet, within this very existence. Life—understood broadly and free from sorrow and pain—is full of wonder when seen through the lens of the present. Were we able to avoid physical pain and ultimately death, as the *xian* do, an immortal life on the earth is not only palatable but desirable.

Given this perspective, we are encouraged to dampen our desires (not extinguish them) and find contentment back at our roots, returning to Dao itself. As the *Daode jing* says,

> Take emptiness to the limit;
> Maintain tranquility in the center.
> The ten thousand things—side-by-side they arise;
> And by this I see their return.
>
> Things [come forth] in great numbers;
> Each one returns to its root.
> This is called tranquility.
> "Tranquility"—This means to return to your fate.
> To return to your fate is to be constant;
> To know the constant is to be wise.

Not to know the constant is to be reckless and wild;
If you're reckless and wild, your actions will lead to misfortune.

To know the constant is to be all-embracing;
To be all-embracing is to be impartial;
To be impartial is to be kingly;
To be kingly is to be [like] heaven;
To be [like] heaven is to be [one with] Dao;

If you're [one with] Dao,
To the end of your days, you'll suffer no harm.
(ch. 16; based on Henricks 1989, 68)

Scholars and practitioners disagree about whether the final line here is meant to refer to immortality (I tend to believe it does). Irrespective of that, however, the chapter illustrates the ultimate goal of returning to the source, the root, which is Dao. Indeed, the very movement of Dao is itself reversal (ch. 40). While we may, in the words of Ge Hong, have the greatest amount of intelligence among all creatures, we are nevertheless nothing more than a single piece of the total. We should use our intelligence to return to the root, not create more distance between ourselves and the Way. Establish and fulfill our desires properly, finding contentment in that childlike state of moment-to-moment engagement with the world.

When the world has the Way,
fleet-footed horses are used to haul dung.
When the world is without the Way,
war horses are raised in the suburbs.

The greatest misfortune is not to know contentment.
The worst calamity is the desire to acquire.
And so those who know the contentment of contentment
are always content. (ch. 46; see Ivanhoe 2003)

Etymologically, the English word "contentment" derives from the Latin *contentus*, meaning "held together" or "intact, whole." Originally used to describe objects that hold things, it eventually evolved into an affective human state, describing one who feels complete, with no desires beyond themselves. To be content, therefore, poses the question, "How whole do you feel as a human being?" In our context, "How close is your relation to Dao?" If content, your desire is managed, as is your desire to desire, that which fosters boredom according to Tolstoy. Hence, as the *Daode jing* says,

Those constantly without desires
by this means will perceive [Dao's] subtlety;
Those constantly with desires
by this means will see only that which they yearn for and seek.
(ch. 1; Henricks 1989, 53)

Manifest plainness and embrace the genuine;
Lessen self-interest and make few your desires;
Eliminate learning and have no undue concerns.
(ch. 19, Henricks 1989, 71)

Desires are not inherently evil, but must be properly controlled and utilized correctly, as we saw above with Pengzu. Yes, the world bombards us with the five colors, the five tones, the five flavors, and so on. However, to cut oneself off from them leads to an artificial life, one devoid of meaning and value. Controlling desire, rather than being controlled by it, allows us to appreciate the joys and amusements of life as they arrive from moment to moment, always fresh, always new, always engaging. How could such a life be boring, whether it be for 337 years or forever?

I conclude with two final examples, neither from immortals' hagiogra-phies. Yet they put an exclamation point on the need to enjoy life, engage the granularity of existence (or better insistence) within nature, and take an optimistic, as opposed to curmudgeonly, outlook on an immortal life

The first comes from the *Zhuangzi*, recounting various imperial attempts to "Give Away the Throne."

> Shun tried to cede the empire to Shan Quan, but Shan Quan said, "I stand in the midst of space and time. Winter days I dress in skins and furs, summer days, in vine-cloth and hemp. In spring I plow and plant—this gives my body the labor and exercise it needs; in fall I harvest and store away— this gives my form the leisure and sustenance it needs.
>
> "When the sun comes up, I work; when the sun goes down, I rest. I wander free and easy between heaven and earth, and my mind has found all that it could wish for. What use would I have for the empire? What a pity that you don't understand me!"
>
> In the end he would not accept, but went away, entering deep into the mountains, and no one ever knew where he had gone. (ch. 28; Watson 1968, 309-10)

The second example is Buddhist-based and comes via James Heisig. He emphasizes that contentment within sufficiency can be found in numerous places in the Buddhist canon. For our purposes, the most relevant instance appears in an early 5th-century text, entitled *Sutra on the Final Instruction of the Buddha*. It has,

> Knowing how much is enough offers a comfortable, secluded spot. . . For one who can never have enough, wealth is still poverty; for one who knows what it is to have enough, there is wealth even in poverty. One who does not know how much is enough is forever pulled this way and that by the desires of the senses; one who knows finds final consolation. (2013, 117-18)

In such a manner, by diminishing the constant drive of desire that keeps us looking for that next fresh experience and produces boredom as we wait, perhaps we can positively envision a Daoist physical earthbound immortality—one that would have us echoing the words of Ge Hong: "Once one's immortality has been confirmed, one is never again concerned about the fleeting of time. If one should return temporarily to wander on earth or in the famous mountains, what would there be to concern oneself about?"

Bibliography

Bokenkamp, Stphen R. 2021. *A Fourth-Century Daoist Family: The Zhen'gao, or Declarations of the Perfected Volume 1*. Berkeley: University of California Press.

Bortolotti, Lisa, and Yujin Nagasawa. 2009. "Immortality without Boredom." *Ratio* XXII:261-77.

Burley, Mikel. 2009. "Immortality and Meaning: Reflections on the Makropulos Debate." *Philosophy* 84:529-47.

Capek, Karel. 1999 [1922]. *The Makropulos Case.* In *His Four Plays*, translated by Peter Majer and Cathy Porter. London: Methuen.

Campany, Robert Ford. 2002. *To Live as Long as Heaven and Earth: A Translation and Study of Ge Hong's Traditions of Divine Transcendents*. Berkeley: University of California Press.

Cave, Stephen. 2012. *Immorality: The Quest to Live Forever and How It Drives Civilization*. New York: Crown.

Chappell, Timothy. 2007. "Infinity Goes on Trial: Must Immortality be Meaningless"? *European Journal of Philosophy* 17.1:30-44.

Cross, Ted. 2014. *The Immortality Game*. Breakwater Harbor, Del.: Breakwater Harbor Books.

DeWoskin, Kenneth J. 1990. "Xian Descended: Narrating Xian Among Mortals." *Taoist Resources* 2.2:70-86.

Eliade, Mircea. 1959. *The Sacred and the Profane: The Nature of Religion*. Translated by William R. Trask. New York: Harvest Books.

Fischer, John Martin. 2020. *Death, Immortality, and Meaning in Life*. New York: Oxford University Press.

Frankl, V. 1957. *The Doctor and the Soul*. New York: Alfred Knopf.

Gardiner, Michael E., and Julian Jason Haladyn, eds. 2019. *Boredom Studies Reader: Frameworks and Perspectives*. London: Routledge.

Gontcharova, Natalie. 2021. "Congratulations, Our Sense of Time Has Been off for a Year Now." https://refinery29.com/en-us/2021/03/ 10332829/why-does-time-feel-weird-covid-pandemic. Accessed March 25, 2021.

Heidegger, Martin. 1995. *The Fundamental Concepts of Metaphysics: World, Finitude, Solitude*. Studies in Continental Thought. Bloomington: Indiana University Press.

Heisig, James W. 2013. *Nothingness and Desire: An East-West Philosophical Antiphony*. Honolulu: University of Hawaii Press.

Henricks, Robert G. 1989. *Te-Tao Ching: A New Translation Based on the Recently Discovered Ma-wang-tui Texts*. New York: Ballantine Books.

Huang, Shih-shan Susan. 2012. *Picturing the True Form: Daoist Visual Culture in Traditional China*. Cambridge, Mass: Harvard University Press.

Huang, Xinkai. 2011. "To Become Immortal: Chinese Fantasy Literature Online." *Intercultural Communication Studies* XX.2:119-30.

Ivanhoe, P. J. 2003. *The Daodejing of Laozi*. Indianapolis: Hackett Publishing.

James, Williams. 1890. *The Principles of Psychology*. Public domain. https://archive.org/details/theprinciplesofp01jameuoft/page/n5/mode /2up. Accessed March 15, 2021.

Kagan, Shelly. 2012. *Death*. New Haven, Conn.: Yale University Press.

_____. 2021. "Death." PHIL 176. Open Yale Course. https://oyc.yale.edu/NODE /196. Accessed March 30, 2021.

Kirkland, Russell. 2004. *Taoism: The Enduring Tradition*. New York: Routledge.

Kirkova, Zornica. 2016. *Roaming into the Beyond: Representations of Xian Immortality in Early Medieval Chinese Verse*. Leiden: Brill.

Lai, Chi-Tim. 1998. "Ko Hung's Discourse on Hsien-Immortality: A Taoist Configuration of an Alternate Ideal Self-Identity." *Numen* 45:183-220.

Lien, Fontaine. 2018. "Stigmas and Rewards of Immortality in Taoist and Gothic Literary Traditions." *Pacific Coast Philology* 53.1:68-91.

Magary, Drew. 2011. *The Postmortal*. New York: Penguin Books.

Miller, James. 2017. *China's Green Religion: Daoism and the Quest for a Sustainable Future*. New York: Columbia University Press.

Mulvey-Roberts, Marie. 2010. *Gothic Immortals: The Fiction of the Brotherhood of the Rosy Cross*. New York: Routledge.

Phillips, Adam. 1993. *On Kissing, Tickling, and Being Bored: Psychoanalytic Essays on the Unexamined Life*. Cambridge, Mass.: Harvard University Press.

Pregadio, Fabrizio. 2018. "Which is the Daoist Immortal Body?" *Micrologus* XXVI: 385-407.

______. 2020. "Seeking Immortality in Ge Hong's *Baopuzi Neipian*." In *Dao Companion to Xuanxue 玄學 (Neo-Daoism)*, edited by David Chai, 427-56. New York: Springer.

Robbins, Tom. 1990. *Jitterbug Perfume*. New York: Bantam.

Robinet, Isabelle. 1997. *Taoism: Growth of a Religion*. Translated by Phyllis Brooks. Stanford: Stanford University Press.

Roth, Harold D. 1999. *Original Tao: Inward Training and the Foundations of Taoist Mysticism*. New York: Columbia University Press.

Russell, Bertrand. 1930. *The Conquest of Happiness*. London: George Allen & Unwin.

Scheffler, Samuel. 2015. *Death and the Afterlife*. Edited by Niko Kolodny. New York: Oxford University Press.

Schipper, Kristofer M. 1993. *The Taoist Body*. Translated by Karen C. Duval. Berkeley: University of California Press.

Shelley, Mary. 1833. "Mortal Immortal." https://www.owleyes.org/text/the-mortal-immortal/read/the-mortal-immortal#root-2. Accessed January 13, 2021.

Simic, Charles. 2007. "To Boredom." *The New Yorker*, December 10.

Spacks, Patricia Meyer. 1995. *Boredom: The Literary History of a State of Mind*. Chicago: University of Chicago Press.

Svendsen, Lars. 2005. *A Philosophy of Boredom*. Translated by John Irons. London: Reaktion Books.

Swift. Jonathan. 2003 [1726]. *Gulliver's Travels*. New York: Barnes and Noble.

Talbot, Margaret. 2020. "What Does Boredom Do to Us—and for Us?" *The New Yorker*, August 20. https://www/newyorker.com/culture/annals-of-inquiry/what-does-boredom-do-to-us-and-for-us. Accessed January 5, 2021.

Thagard, Paul. 2020. "What is COVID Fatigue?" *Psychology Today*, November 12. https://www.psychologytoday.com/us/blog/hot-thought/202011/what-is-covid-fatigue. Accessed January 15, 2021.

Thompson, Jon W. 2019. "'Intimations of Immortality': A Response to Bernard Williams." *Religious Studies* 55:245-60.

Thoreau, Henry David. 1992 [1854]. *Walden and Other Writings.* New York: Barnes and Noble Books.

"Time Enough at Last." November 20, 1959. *The Twilight Zone,* created by Rod Serling. Season 1, episode 8. CBS Productions.

"Tithonus." January 24, 1999. *The X-Files,* created by Chris Carter. Season 6, Episode 10. Ten Thirteen Productions, 20th Century Fox.

Tolstoy, Leo. 2004. *Anna Karenina.* Translated by Richard Pevear and Larissa Volokhonsky. New York: Penguin Classics.

Ware, James R. 1966. *Alchemy, Medicine, Religion in the China of A.D. 320: the Nei p'ien of Ko Hung (Pao-p'u tzu).* Cambridge, Mass.: MIT Press.

Williams, Bernard. 2010. "The Makropulos Case: Reflections on the Tedium of Immortality." In *Life, Death, and Meaning: Key Philosophical Readings on the Big Questions,* edited by David Benatar, 345-62. Lanham, MD: Rowman & Littlefield Publishers.

Wittmann, Marc. 2020. "The Felt Passage of Time During the Pandemic." *Psychology Today,* November 6. https://www.psychologytoday.com/us/blog/sense-time/202011/the-felt-passage-time-during-the-pandemic. Accessed December 4, 2020.

Wordsworth, Henry. 2011. *Ode on Immortality and Other Poems.* Edinburgh: W. P. Nimmo, Hayt and Mitchell. Also at https://poets.org/poem/ode-intimations-immor-tality-recollections-early-childhood. Accessed February 3, 2021.

Yü, Ying-shih. 1964. "Life and Immortality in the Mind of Han China." *Harvard Journal of Asiatic Studies* 25:80-122.

Contributors

Jeffrey W. Dippmann, Ph. D., is professor and department chair in Religious Studies and Philosophy as well as Director of Asian Studies at Central Washington University, where he is also University Distinguished Professor in Service. His main research interests lie in early Daoist thought and comparative philosophy/religion. In 2001, he co-edited *Riding the Wind with Liezi: New Perspectives on the Daoist Classic* with Ronnie Littlejohn (Belmont University).

Andrej Fech, Ph.D., is assistant professor at the Department of Chinese Language and Literature at Hong Kong Baptist University. His research focuses on early Chinese intellectual history, particularly Daoist thought, excavated manuscripts, and comparative philosophy.

Wujun Ke is a doctoral candidate in the Department of Comparative Literature at the University of California at Irvine. She earned her B. A. in Comparative Literature at the University of Chicago in 2013. She specializes in contemporary Sinophone cinemas, theories of temporality, and documentary ethics.

Livia Kohn, Ph.D., is professor emerita of Religion and East Asian Studies at Boston University. The author or editor of close to sixty books (including the annual *Journal of Daoist Studies*), she spent ten years in Kyoto doing research. She now serves as the executive editor of Three Pines Press, runs international conferences and workshops, and guides study tours to Japan.

Patrick Laude, Ph.D., joined the faculty of Georgetown University in Washington, DC in 1991 and is currently professor of Religious Studies at Georgetown in Qatar. He has authored a dozen books on metaphysics and mysticism including *Shimmering Mirrors: Reality and Appearance in Contemplative Metaphysics East and West* (2017), *Divine Play, Sacred Laughter and Spiritual Understanding* (2005), and *Surrendering to the Self: Ramana Maharshi's Message for the Present* (2021).

Joseph L. Pratt, J. D., taught law and researched Chinese philosophy and logic from 2011 to 2020 at Peking University. During that time, he also practiced taiji quan under the guidance of Master Yang Songquan. He continues to research and write on Daoism as a holistic account of reality.

Nada M. Sekulic, Ph. D., is a Serbian anthropologist, philosopher, and feminist at the Faculty of Philosophy, University of Belgrade. The author of several books (*On the End of Anthropology, The Hidden War, The Culture of Giving Birth*), she has published dozens of essays in various academic journals. She is also a certified instructor of Wudang-style taiji quan.

Qi Song 宋琦, Ph. D., serves as lecturer at Jiangxi University of Science and Technology and currently does research at the International Research Center for Japanese Studies in Kyoto, where she also earned her degree. Her academic focus is on the interaction of the three teachings in Tokugawa Japan.

Steve Taylor, Ph.D., is senior lecturer in psychology at Leeds Beckett University and current chair of the Transpersonal Psychology Section of the British Psychological Society. Among his thirteen books, published in twenty languages, are *The Clear Light, Waking from Sleep, Out of the Darkness, The Calm Center*, and *Spiritual Science*. He regularly publishes academic papers as well as articles in the popular media. His website is www.stevenmtaylor.com.

Mercedes Valmisa, Ph.D., is assistant professor of Philosophy at Gettysburg College. Her book, *Adapting: A Chinese Philosophy of Action* (2021), explores an extraordinary strategy for efficacious relational action devised by classical Chinese philosophers in order to account for the interdependent and embedded character of human agency. She specializes in Chinese, Asian, and comparative philosophy.

Juan Zhao 赵娟, Ph.D., is associate professor at the School of Humanities, Beijing Sport University. Her research focuses on Chinese classical aesthetics and art history in trans-culture. She has published close to thirty articles and two translated works in related fields.

Index